STUDENT'S STUDY MANUAL

TO ACCOMPANY VAN WYNSBERGHE, NOBACK, AND CAROLA

Human Anatomy & Physiology

STUDENT'S STUDY MANUAL

TO ACCOMPANY VAN WYNSBERGHE, NOBACK, AND CAROLA

Human Anatomy & Physiology

THIRD EDITION

Gail R. Patt

Boston University

McGraw-Hill, Inc.

New York   St. Louis   San Francisco   Auckland   Bogotá   Caracas   Lisbon   London   Madrid
Mexico City   Milan   Montreal   New Delhi   San Juan   Singapore   Sydney   Tokyo   Toronto

STUDENT'S STUDY MANUAL
TO ACCOMPANY VAN WYNSBERGHE, NOBACK, AND CAROLA:
HUMAN ANATOMY & PHYSIOLOGY

 This book is printed on recycled paper.

1 2 3 4 5 6 7 8 9 0 SEM SEM 9 0 9 8 7 6 5 4

ISBN 0-07-011174-X

The editor was Kathi M. Prancan;
the production supervisor was Diane Ficarra.
Semline, Inc., was printer and binder.

CONTENTS

PREFACE

A course in human anatomy and physiology is often the first step in achieving the goal of a career in the health sciences. Even though they anticipate that the course will be difficult, most students look forward to the experience because essentially all the material is meaningful, concrete, interesting, and relevant. However, despite a prior understanding that this course will not be lightweight, most students are genuinely surprised at just how hard they must work to master the breadth and depth of the material. While some will find areas of conceptual difficulty, and others will breeze through the most intricate of physiological explanations, all will soon come to understand that the volume of material in a good anatomy and physiology course precludes cramming or lack of attention to detail. Anatomy and physiology are best learned by active, repetitive reading, by active manipulation of the material, and by active teaching of the material to one's peers. Although several readings of each chapter and a good set of lecture notes will be essential, the key word should be "active." Reading without constant questioning, rephrasing, and reframing of the concepts will result in a minimal level of preparation or understanding. Similarly, you can prepare hundreds of flash cards, each containing a key word and its definition, but if the definition is simply copied from the textbook, the effort will merely be an inefficient use of precious time. Moreover, the effort must be continual. It is the wise student who realizes that far more can be accomplished with shorter daily study sessions than with one long weekly session. Success in anatomy and physiology resembles completion of a marathon rather than running a hundred-yard dash.

This manual was written to help students become active, repetitive learners—in effect to learn how to learn. When used properly, in combination with text and lectures, it will guide you toward mastery of the subject.

Contents of the Study Guide Chapters

Each chapter in the manual begins with a lengthy section entitled "Active Reading" that closely follows the organization of material in your text. This is followed by "Key Terms," an alphabetical list of the important new terms introduced in the chapter, along with the page on which they are first used and defined. Next comes "Post Test," which simulates an actual test of the material.

The final two sections in each chapter are "Integrative Thinking" and "Your Turn." The purpose of both sections is the conversion of a mass of separately learned facts into a meaningful web of knowledge that is much more likely to be retained. The problems and puzzles posed in "Integrative Thinking" encourage you to make the same kind of knowledge connections you will have to make later on in both your professional and personal lives. Some pose clinical problems while others ask you to add information presented in this chapter to that found in previous chapters. "Your Turn" encourages you to shift your focus from that of being a solitary student to that of being a teacher. Join with a few friends to form a study group. It does not matter whether some students in the study group are high achievers or not. The only important criterion you should use when asking friends to join your group is whether they are serious about learning this material. The best way to learn is to teach; by following the instructions in "Your Turn," you will begin to make that transition.

How to Use the Manual

Just as each section in each chapter has a particular purpose, each has its own best time, place, and method of use. The first step, even before cracking open a chapter in the manual, is a quick read through the assigned chapter in your textbook. The objective of the reading is familiarity with new vocabulary and an overview of the material. Is it primarily anatomical? Physiological? A combination of both? Does it make sense? Does it fit in with what you have already learned? Ideally this reading should be done before your instructor lectures on the material because it is difficult to understand new concepts while simultaneously trying to figure out how to spell a word.

After the appropriate lecture, it is time to do a second reading of the chapter. This reading should be slow and deliberate, and it should include doing the "Active Readings" and "Key Terms" sections of the study guide. Both books should be open and in simultaneous use. After you read each section of the textbook, put the textbook aside and complete the proper section in "Active Reading." It will be easy for you to match sections in the textbook with those in the study guide because each title of each section in "Active Reading" is followed by the proper textbook page reference. In all probability, you will require several study sessions in order to complete one chapter, but if done properly, "Active Learning" will force you to stop, think, and write. It is a way to put your brain in the loop! Next, go through the key terms, and define each in your own words. Use each term in several sentences in order to fix the concept in your long-term memory bank.

After completing "Active Learning" and "Key Terms," it is time to take the "Post Test." Try to put yourself in an exam situation. Sit at a desk. Turn off the radio. Don't have a snack. Don't look up the answers at the back of the manual until the entire test has been completed. Time yourself. Each "Post Test" should last about 30 minutes. When you have finished, check your answers, and review the textbook in places where you made errors.

Both the "Active Learning" and "Post Test" sections of the manual contain illustrations, many of which were specially drawn for the manual. Anatomy is a visual science that cannot be entirely learned with words alone. Label each of the figures carefully; the time spent will be well worth it.

When you are feeling refreshed, and preferably when you are in your study group, tackle the last two sections of the chapter. Answer the "Integrative Thinking" questions out loud. Be complete. State all the facts upon which you have based your answer. If you are doing this in your study group, the group should act as a friendly but critical chorus. Try to poke holes in each other's logic. Try to come up with several answers and then defend each in turn. After you have exhausted all possibilities, consult the answer section at the end of the manual.

Finally, move on to "Your Turn." These exercises have been designed as a group activity. They can be done individually, but then you will lose much of the dynamism that makes learning fun. If your group has access to a blackboard, so much the better. If no blackboard is available, be sure to bring a pad of paper. The exercises get progressively harder as you move through the book, but even the introductory exercises will help you learn.

If you have faithfully and energetically followed all of these instructions, you will be well on your way to storing the material in your long-term memory bank. You will have mastered the terminology and the concepts well before the exam, and the need for ineffective cramming will be gone.

When final exam time rolls around, you will find the "Active Learning" section once again to be invaluable. Many students find that the concentration of final exams in several courses prevents a complete and thorough rereading of the textbook. In a pinch "Active Learning" can serve as an outline of the textbook, and in combination with a good set of lecture notes, can provide the focus for study.

Several people were essential to the completion of this study guide. Donald I. Patt prepared the majority of the illustrations and helped with the ideas and development of the problems posed for "Integrative Thinking" and "Your Turn." Anthony G. Patt played the role of the novice student, was instrumental in the development and logic of "Active Learning," designed the text, and prepared the camera-ready copy. Katherine Hedges of Indiana University Northwest provided feedback on preliminary chapters. None of the manual would have been possible without the editorial efforts of Kathi M. Prancan, Pamela Wirt, and Sharon Geary.

Gail R. Patt

STUDENT'S STUDY MANUAL

TO ACCOMPANY VAN WYNSBERGHE, NOBACK, AND CAROLA

Human Anatomy & Physiology

1 Introduction to Anatomy and Physiology

Active Reading

I. What Are Anatomy and Physiology? (3)

1. What is anatomy? _____

2. Define the major specialties of anatomy: _____

    a. gross anatomy _____

    b. microscopic anatomy _____

    c. embryological anatomy _____

    d. developmental anatomy _____

    e. radiographic anatomy _____

3. What are regional and systemic anatomy? _____

4. What is physiology? _____

5. Why are anatomy and physiology studied together? _____

II. Homeostasis: Coordination Creates Stability (3)

A primary and fundamental example of the relationship between anatomy (structure) and physiology (function) is homeostasis.

1. What is homeostasis? _____

2. How is homeostasis maintained? _____

3. Name four functions of the body that are controlled by homeostatic mechanisms.

    a. _____

    b. _____

    c. _____

    d. _____

4. What is extracellular fluid, and what role does it play in maintaining homeostasis? _____
   _____

5. What are negative feedback and positive feedback?
   a. Negative feedback _____
   b. Positive feedback _____

6. Under what circumstances do feedback mechanisms come into play? _____

7. Describe an example of negative feedback in the body. _____

8. What is meant by stress? _____

9. Name some common stressors. _____

10. What is the relationship between stress, stressors, and homeostasis? _____

III. From Atom to Organism: Structural Levels of the Body (7)

1. In the spaces below, rearrange the following items in ascending order from the smallest and simplest to the biggest and most complex. In jumbled order, these items are: compounds, tissues, atoms, cells, systems, molecules, organs, and organism.

   a. _____
   c. _____
   e. _____
   g. _____
   b. _____
   d. _____
   f. _____

2. Complete each of the definitions below with one of the following items.

Items: (a) atom, (b) element, (c) molecule, (d) compound, (e) cell, (f) tissue, (g) organ, (h) system, (i) organism

Definitions:

   a. A collection of similar cells that perform a specific function is a(n) _____ .
   b. The basic unit of all matter is the _____ .
   c. An integrated collection of two or more kinds of tissue that work together for a specific function is a(n) _____ .
   d. A specific kind of atom is known as a(n) _____ .
   e. A molecule consisting of more than one kind of atom is known as a(n) _____ .
   f. A group of organs that serve a common, major function is known as a(n) _____ .
   g. The smallest, self-contained unit of living substance is the _____ .
   h. A dynamic assemblage of all described above is a(n) _____ .

3. Complete the table below by naming or briefly describing the functions of the various tissues named.

Tissue	Functions
Epithelial	
Connective	
Muscle	
Nervous	

IV. Body Systems (10)

1. Name two organs that, together with a number of other organs that we'll learn about later, are part of the digestive system.

   a. _____

   b. _____

2. What is the name of the system that covers the body, helps to regulate body temperature, and contains sensory receptors? _____

3. What system includes all the bones of the body? _____

4. What system includes the numerous glands that secrete the hormones that help regulate many of our body functions?_____

5. What is the function of the respiratory system? _____

6. Name the system that is concerned with each of the following functions.

   a. reproduction, heredity _____

   b. moving the body and its parts _____

   c. digestion and absorption of food _____

   d. circulating blood throughout the body _____

   e. eliminating metabolic wastes _____

   f. receiving and interpreting information from sensory organs _____

   g. returning excess fluid and proteins to the blood and serving as part of the immune system _____

7. Identify the system that includes the organs named.

   a. pharynx, larynx, and lungs _____

   b. mouth, liver, and pancreas _____

   c. tonsils and spleen _____

   d. kidney, ureter, and urethra _____

   e. heart and blood vessels _____

   f. pituitary, adrenals, and thyroid _____

   g. uterus, vagina, and ovary _____

   h. prostate, scrotum, and penis _____

    i.   trachea and diaphragm _____

    j.   testes, pancreas, and thymus _____

V. Anatomical Terminology (14)

Match each of the directional descriptions below with the appropriate anatomical equivalent in the following list. Answer in the space provided.

____ 1. Toward the back of the body in its anatomical position

____ 2. A lower or downward position

____ 3. A higher or upward position

____ 4. Toward the front of the body in its anatomical position

____ 5. Nearer the trunk of the body or, in the case of a limb, toward the attached end

a. superior
b. inferior
c. anterior or ventral
d. posterior or dorsal
e. proximal

Match the anatomical terms below and the descriptions that follow:

____ 6. Farther from or away from the midline of the body

____ 7. Nearer the surface of the body

____ 8. Nearer or toward midline of the body

____ 9. Farther from the trunk or, in the case of a limb, away from the attached end

____ 10. Farther from the surface of the body

a. medial
b. lateral
c. distal
d. superficial
e. deep

11. Complete the following descriptions by writing the appropriate anatomical terms or directions in the spaces provided: The sole or bottom of the foot is called the _____ surface, and its upper surface is called the _____ surface. The palm of the hand is the _____ surface, and the back of the hand is its _____ surface. The head is _____ to the neck. The hand is _____ to the wrist, and the wrist is _____ to the hand. The breastbone (sternum) is _____ to the backbone (spinal column), and the heel is _____ to the toes. The ribs are _____ to the lungs.

12. Give the technical (anatomical) name for each of the following lay terms:

    a.   nape of the neck _____

    b.   loin _____

    c.   back of the elbow _____

    d.   buttock_____

    e.   calf of the leg _____

    f.   shoulder _____

    g.   back of the knee _____

    h.   heel of the foot _____

13. Label the two figures below where indicated.

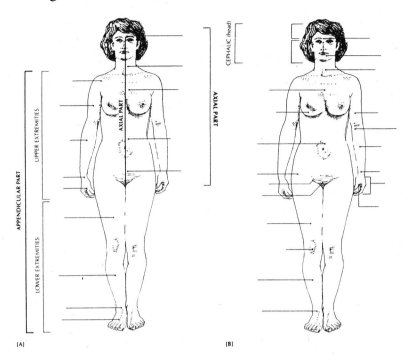

14. Name the nine subdivisions of the abdominal region of the body as illustrated below.

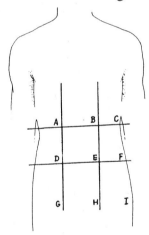

a. _____

b. _____

c. _____

d. _____

e. _____

f. _____

g. _____

h. _____

i. _____

15. Label the body planes shown in the figure below.

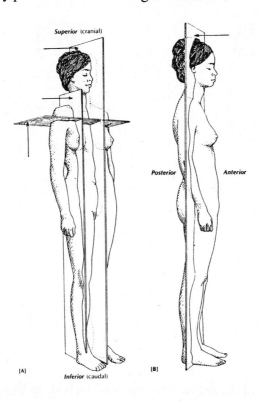

VI. Body Cavities and Membranes (22)

1. Label the main body cavities represented in the figure below. They are the dorsal body cavity, including the cranial and spinal cavities, and the ventral body cavity, which includes the thoracic and abdominopelvic cavities. The thoracic cavity contains the pleural cavities and the mediastinum, which in turn contains another cavity, the pericardial cavity. The abdominopelvic cavity is divided into the abdominal cavity superiorly, the pelvic cavity inferiorly, and the inconspicuous peritoneal cavity.

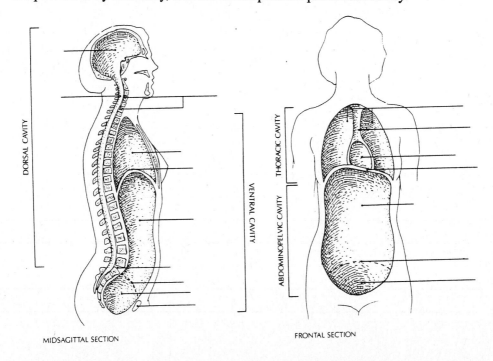

2. What are the three main types of membranes?

    a. _____

    b. _____

    c. _____

3. Of what are these membranes composed? _____

4. What is the function of the body membranes? _____

5. What is the function of the serous fluid of the body membranes? _____

6. (a) What is the distinction between visceral and parietal portions of membranes, and (b) what does the term retroperitoneal refer to?

    a. _____

    b. _____

Key Terms

abdominopelvic 23
anatomical position 14
anatomy 3
anterior 14
appendicular 14
atom 7
axial 14
cavity 22
cell 7
compound 7
distal 14
dorsal 23
external 14
feedback systems 3

frontal 16
homeostasis 3
inferior 15
internal 14
lateral 14
medial 14
membrane 22
midsagittal 16
molecule 7
organ 10
organism 10
palmar 14
parietal 14
peripheral 14

physiology 3
plantar 14
posterior 14
proximal 14
retroperitoneal 23
sagittal 16
superficial 14
superior 14
system 10
tissue 7
transverse 16
ventral 23
visceral 14

Post Test

Multiple Choice

_____ 1. While all body systems are maintaining balanced physiological conditions internally in a constantly changing environment externally, the body is said to be in a state of

    a. apoplexy
    b. dormancy
    c. erratic metabolism
    d. homeostasis
    e. paralysis

_____ 2. Of primary importance in achieving and maintaining the conditions described in the preceding question is

    a. positive feedback
    b. the extracellular fluid
    c. stress
    d. a proper attitude
    e. high levels of blood calcium

____ 3. Also of great importance in controlling the constancy of the internal
environment is (are)

    a. properly functioning feedback mecha-
nisms
    b. healthy kidneys
    c. a sound circulatory system
    d. the respiratory system
    e. all of the above

____ 4. One mechanism by which the body
responds automatically to changes in
the internal environment is

    a. feedback control
    b. homeostasis
    c. process inhibition
    d. synergistic response
    e. both b and c

____ 5. Feedback activity that produces a
response opposite the initiating
stimulus is said to be

    a. positive
    b. negative
    c. ineffective
    d. neutral
    e. alternately positive and negative

____ 6. If blood pressure decreases below
normal, the feedback response that
causes it to return to normal is

    a. positive feedback
    b. negative feedback
    c. regulated by the adrenal glands
    d. alternately positive and negative
    e. temporarily inactivated

____ 7. Factors that cause homeostatic
imbalances in the body are called

    a. perturbations
    b. homeostatic alarms
    c. antistasis factors
    d. stressors
    e. all of the above

____ 8. An example of a chemical compound is

    a. oxygen
    b. helium
    c. carbon
    d. water
    e. all of the above

____ 9. Which of the following is <u>not</u> a
chemical compound?

    a. carbon dioxide
    b. carbohydrate
    c. protein
    d. lipid
    e. sulfur

____ 10. The smallest independent units of
life are called

    a. macromolecules
    b. DNA
    c. cells
    d. viruses
    e. yeasts

Short Answer

11. Define tissue? _____

_____

12. Name the four kinds of tissue, the function of each, and a specific example of where each may be found.

Name	Function	Location

13. In the table below, name the seven systems of the body and cite the major function or functions of each.

System	Functions

14. Which of the figures below illustrates the anatomical position?

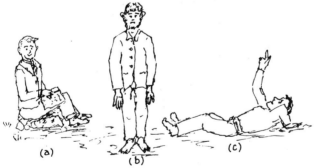

(a)   (b)   (c)

15. What is figure (a) doing?_____

Matching

____ 16. Left lumbar region
____ 17. Right iliac region
____ 18. Hypogastric region
____ 19. Epigastric region
____ 20. Umbilical region
____ 21. Left hypochondriac region

a. upper abdomen
b. middle abdomen
c. lower abdomen
d. sacral quadrant

___ 22. Heart
___ 23. Lungs
___ 24. Urinary bladder
___ 25. Spleen

a. pleural cavity
b. abdominal cavity
c. pelvic cavity
d. mediastinum

___ 26. Divides the body into left and right halves.
___ 27. Divides the body into anterior and posterior sections.
___ 28. A coronal plane.
___ 29. Divides the body into unequal left and right sections.
___ 30. The horizontal plane.

a. frontal plane
b. sagittal plane
c. midsagittal plane
d. transverse plane

Identify the principal distinguishing feature or use of each of the following diagnostic techniques.

31. CT scan _____

32. PET scan _____

33. Sonography _____ _____

34. MRI _____

35. Thermography _____

Integrative Thinking

1. A routine chest x-ray discloses the presence of a small nodule on the right, outer surface of the right lung of a person. The finding is confirmed by CT scan. In order to determine whether or not the nodule is cancerous, a biopsy is indicated. As the chest surgeon, you decide to use thoracoscopy to obtain a sample of the growth, perhaps even to remove it entirely, since it is so small. Describe how you will go about the operation, including naming the tissues and/or organs though which you will have to go.

2. When a pregnancy reaches term, the hypothalamus signals the release of the hormone oxytocin by the posterior pituitary into the general circulation. When it reaches the uterus, this hormone causes the uterine muscles to contract, and active labor is initiated. The pressure thus applied by the contractions stimulates receptors in the uterine wall to send reflex messages to the hypothalamus, stimulating it to secrete more oxytocin. As this progresses, the contractions become stronger, and more oxytocin is secreted and the contractions become even stronger. Of course it all ends, happily, when the baby is born, but in the meantime an interesting feedback scenario is evident. (a) Is the main player positive or negative feedback, and (b) how is the maintenance of homeostasis affected?

Your Turn

Go back to the Preface and read again what this section is all about. An example of one of the kinds of things you can do is this: Assemble a group of terms or phrases in random or jumbled order, and then compose a paragraph that will have a number of blank spaces into which these terms will fit. Here is an example:

List of Terms: superior, mediastinum, liver, lungs, organ, pleural cavities, abdominal cavity, pericardial cavity, thoracic cavity

Paragraph: The heart is an _____ that is located in the _____, which lies in the _____ between the two _____ in which are situated the two _____. All of these structures occupy the _____, which is _____ to the _____, in which the _____ , among several other organs, is situated.

Now that you have an idea of how to go about this, here is a list of terms for you. See what you can compose. As you get better at it, they'll get harder!

List of Terms: medial, proximal, superficial, peripheral, parietal, distal, visceral, lateral.

2 Essentials of Body Chemistry

Active Reading

I. Matter, Elements, and Atoms (31)

1. What is the difference between *matter, mass,* and *weight?* _____

2. Distinguish between *atom, molecule,* and *compound.* _____

3. Complete the table below.

TABLE 2.1 Some Chemical Elements Essential to Body Function

Element Symbol	Percent of body weight	Significance to Human Body
O		
C		
H		
N		
Ca		
P		
K		
S		
Cl		
Na		
Mg		
I		
Fe		

4. What are the basic particles that make up atomic structure, and what is their electromagnetic charge? _____

5. What is the difference between atomic weight and atomic number? _____

6. Are both atomic weight and atomic number always the same for a given element? _____. When do they differ? _____

7. What is a radioisotope? _____

8. What are the three major categories of emissions from radioisotopes?

   a. _____

   b. _____

   c. _____

9. How are radioisotopes used as diagnostic tools? _____

10. What are electron shells, and how are they involved in chemical reactions? _____

11. In the space below, make a diagram of (a) a hydrogen atom and (b) a carbon atom, showing electron shells, electrons, and nuclei.

a.   b.

II. How Atoms Combine (34)

____ 1. When atoms interact chemically to
         form molecules, the atoms are held
         together by electrical forces called

   a. ionic bonds
   b. molecular bonds
   c. hydrogen bonds
   d. chemical bonds
   e. covalent bonds

____ 2. Ionic bonds are formed when a
         group of atoms gives up electrons to
         another group of atoms, which
         receives them in like number. In the
         process, both kinds of atoms become

   a. electrically neutral
   b. positively charged
   c. negatively charged
   d. oppositely charged
   e. similarly charged

____ 3. Na^+, Ca^{2+}, and Fe^{3+} are all

   a. cations
   b. isotopes
   c. electron rich
   d. anions
   e. molecules

____ 4. When atoms share outer shell
         electrons with other atoms, the
         chemical bond that is formed is
         called

   a. atomic
   b. ionic
   c. oxidative
   d. hydrogen
   e. covalent

___ 5. Covalent bonds are

        a. weak
        b. strong
        c. polar
        d. bipolar
        e. unstable

6. What is special about hydrogen bonds? _____

7. What special properties do hydrogen bonds give to water? _____

III. Chemical Reactions (38)

1. Define the following terms.

    a. metabolism _____

    b. anabolism _____

    c. catabolism _____

    d. Synthesis _____

    e. reactant _____

    f. Reversible reaction _____

    g. oxidation reaction _____

    h. reduction reaction _____

    i. oxidation-Reduction reaction _____

    j. hydrolysis _____

2. What kind of reaction is the following? $C_6H_{12}O_6 + C_6H_{12}O_6 \rightarrow C_{12}H_{22}O_{11} + H_2O$ _____

3. What is the reverse of the preceding reaction called? _____

IV. Water (39)

1. Complete the following sentences by inserting in the spaces provided the appropriate terms from this list: positive, negative, asymmetric, polar molecule, hydrogen bonds. Water is an _____ molecule, and because of this and the resultant distribution of _____ and _____ charges, it is said to be a _____. Water can form _____ with other water molecules and with a variety of other compounds.

2. Name the four major properties of water that make it a very special molecule. (In fact, life as we know it would not be possible without the two wonder molecules: water and carbon.)

    a. _____

    b. _____

    c. _____

    d. _____

V. Acids, Bases, Salts, and Buffers (41)

1. What is an electrolyte? _____

2. What is an acid? _____

3. What is a base? _____

4. What is a salt? _____

5. What is the following equation called? $HCl + NaOH \rightarrow NaCl + H_2O$ _____

6. Identify each of the four molecules in the equation above and state which are reactants and which are products.

   a. _____

   b. _____

   c. _____

   d. _____

7. Why do lemons taste sour? _____

8. Why is soap slippery? _____

9. Compare hydrochloric acid and acetic acid. _____

10. What does pH mean? _____

11. What does it measure? _____

12. What does pH 6.5 signify? _____

13. What does pH 8.0 signify? _____

14. What does pH 7.0 signify? _____

15. What are chemical buffers? _____

16. Why are buffers important to homeostasis? _____

VI. Some Important Organic Compounds (43)

1. Name the four major groups of organic compounds that constitute the chemistry of living matter.

   a. _____

   b. _____

   c. _____

   d. _____

Match the following.

___ 2. Components of all organic compounds

___ 3. Sucrose and maltose

___ 4. Maltose and lactose

___ 5. Glucose, fructose, and galactose

___ 6. Basic formula for carbohydrates in
general

___ 7. $C_{12}H_{22}O_{11}$

___ 8. $(C_6H_{10}O_5)_n$

___ 9. End product of hydrolysis of starch
or of glycogen

___ 10. An end product of dehydration
synthesis of maltose

___ 11. Heparin and cellulose

a. monosaccharides
b. disaccharides
c. polysaccharides
d. CH_2O
e. carbon and hydrogen

12. Following is a list of organic compounds. Which ones are not lipids?

    a. fats _____

    b. glycerol _____

    c. prostaglandins _____

    d. steroids _____

    e. cholesterol _____

    f. triglycerides _____

    g. bile salts _____

    h. ribose _____

    i. sex hormones _____

13. Fatty acids have two regions, a hydrocarbon chain and a chemically reactive group
called a _____

14. A molecule of fat is formed when the three _____ groups in a molecule of glycerol combine with the reactive _____ groups of three fatty acid molecules.

15. The preceding reaction is another example of _____ synthesis.

16. Phospholipids are fatty compounds that contain glycerol, two fatty acid residues, and a
_____ group.

17. Since the two fatty acid chain components of a phospholipid molecule are water in-
soluble, whereas the rest of the molecule is hydrophilic (attracts water), phospholipids are
said to be _____ compounds.

18. Lipids that are composed of three six-sided carbon rings plus a five-sided carbon ring, all
covalently bonded, are called _____. The hormones testosterone and estrogen
are examples of this category of lipid molecules.

19. Name the four elements that are components of all proteins. _____

20. What is the class of proteins that regulate virtually all the chemical processes that take place in the body? _____

21. What are the chemical compounds that are the "building blocks" of all proteins?_____

22. Identify the chemical class that each of the molecules diagrammed below represents.

a.   b.   c.

a. _____

b. _____

c. _____

23. In the space provided below, sketch the structural formula of (a) an amino group, (b) a carboxyl group, and (c) a hydroxyl group.

a. _____

b. _____

c. _____

24. Complete the following dehydration synthesis reaction.

a.   +   b.   ⟶   c.

25. Name and define the four levels of protein structure.

a. _____

b. _____

c. _____

d. _____

26. What is the primary function of enzymes? _____

27. Why is the *shape* of an enzyme important? _____

28. What is meant by the term *enzyme specificity*? _____

29. How do many enzyme inhibitors work? _____

30. What are the nonprotein components of conjugated proteins called? _____

31. Identify three environmental requirements for enzymes to operate properly. _____

32. What are the molecular components of a nucleic acid? _____

33. What are the three molecular components of a nucleotide?

     a. _____

     b. _____

     c. _____

34. What four nitrogenous bases characterize DNA?

     a. _____

     b. _____

     c. _____

     d. _____

35. What four nitrogenous bases does RNA have?

     a. _____

     b. _____

     c. _____

     d. _____

36. What is the structural difference between purines and pyrimidines? _____

37. What are the functions of the nucleic acids? _____

VII. Energy for Living: ATP (52)

____ 1. The immediate source of energy for most biological activities in the cell is a small organic molecule known as

    a. uracil
    b. adenine
    c. guanine
    d. deoxyribose
    e. adenosine triphosphate

____ 2. This energy comes mainly from the bonds of the last two phosphate groups that, along with a third such group, form a sort of tail on a molecule of

    a. adenine
    b. cholesterol
    c. ribose
    d. guanine triphosphate
    e. adenosine

____ 3. A chemical reaction that results in the release of energy, such as the breaking of ATP's high energy bonds

    a. exergonic
    b. endergonic
    c. hydrolytic
    d. synthesis reaction
    e. fortuitous

____ 4. The covalent bonds of the terminal two phosphate groups of ATP are quite unstable and are known as

    a. energy-rich bonds
    b. reversible bonds
    c. high-energy bonds
    d. both a and c
    e. availability bonds

Key Terms

acid 41
amino acid 47
atomic number 32
atomic weight 32
ATP 52
base 41
bond 36
buffer 42
carbohydrate 43
covalent bond 36
dehydration synthesis 38
deoxyribonucleic acid (DNA) 50
disaccharide 43
dissociation 41

electrolyte 41
electron 32
electron shell 34
element 31
enzyme 48
hydrogen bond 37
hydrolysis 38
ionic bond 34
isotope 32
lipid 44
monosaccharide 43
neutron 32
nucleic acid 50
nucleotide 50

nucleus (atomic) 32
oxidation 38
peptide bond 47
pH scale 42
phospholipid 46
polar molecule 39
polysaccharide 44
protein 46
proton 32
reduction 38
ribonucleic acid (RNA) 50
salt 41
water 39

Post Test

Matching

____ 1. Atoms or molecules that carry a positive or negative charge.

____ 2. Calculated as the sum of the number of protons and neutrons.

____ 3. Formed by the apparent distribution of negative charges outside the atomic nucleus.

____ 4. The number of positively charged articles in the atomic nucleus.

____ 5. Molecules that release protons (hydrogen ions) when dissolved in water.

____ 6. Atoms with the same atomic number but different atomic weights.

____ 7. A substance that releases hydroxyl ions when dissolved in water.

a. atomic number
b. atomic weight
c. isotopes
d. electron shells
e. ions
f. acids
g. bases

Multiple Choice

____ 8. A chemical bond that forms when an electrically charged atom is attracted to an atom having an opposite but equal electrical charge

a. ionic
b. covalent
c. hydrogen
d. electronic
e. equalizing

____ 9. When atoms share electrons to fill their outer electron shell, the resultant chemical bond is called a(n)

a. ionic bond
b. covalent bond
c. hydrogen bond
d. electronic bond
e. neutralizing bond

___ 10. The bond that forms when a co-valently bonded hydrogen atom in a compound acquires a relatively positive charge and becomes at-tracted to a relatively negatively charged atom is called a(n)

    a. ionic bond
    b. covalent bond
    c. hydrogen bond
    d. electronic bond
    e. weak dipole bond

___ 11. A chemical reaction in which particular molecules lose electrons is called a(n) _____ reaction.

    a. hydrolysis
    b. reduction
    c. combination
    d. oxidation
    e. decomposition

___ 12. A chemical reaction in which particular molecules gain electrons is called a(n) _____ reaction.

    a. hydrolysis
    b. reduction
    c. combination
    d. oxidation
    e. decomposition

___ 13. Water is very effective as a

    a. solvent
    b. lubricant
    c. transporter
    d. all of these
    e. temperature regulator

___ 14. The separation of molecules into ions when dissolved in water is called

    a. hydration
    b. decomposition
    c. dehydration
    d. reduction
    e. dissociation

___ 15. A compound (other than water) that forms during a neutralization reac-tion between an acid and a base is a(n)

    a. neutron
    b. dipole
    c. salt
    d. monopole
    e. buffer

16. Complete the reaction below, showing the products of the reaction and representing the bonds clearly.

17. What does the symbol "pH" mean?_____

18. What is neutrality on the pH scale? _____

19. What are chemical buffers? _____

20. What effect do buffers have on homeostasis?_____

21. Identify each of the molecules, A – D, represented in the figure.

A_____ B_____ C_____ D_____

22. Name the atomic groupings or bonds indicated in the figures above.

a. _____

b. _____

c. _____

d. _____

e. _____

f. _____

23. Complete the table.

Chemical Family	Building Block	Polymer Example
Carbohydrate		
Protein		
Nucleic acid		

24. Name the component molecules and atomic groupings that make up ATP.

a. Molecule *a* _____

b. Molecule *b* _____

c. Group(s) _____

25. Proteins that increase the rate of a chemical reaction (catalysts) but are not permanently changed by the reaction are known as _____

Integrative Thinking

1. You have a meal that starts with stir-fried caviar and goes on to baked haddock, broiled swordfish, and a tunafish salad. For dessert you have strawberry shortcake. Your companion takes you to task for selecting such a combination of foods, and you say, "What's so bad about strawberry shortcake?" Your companion shakes her head, sighs, and says, "It's a wonder you don't turn into a fish!" Now, there's a thought! Why don't you turn into a fish, or at least develop fishy muscles, after consuming all that fish protein?

2. Many people have a condition known as lactose intolerance, which renders them unable to eat dairy products without suffering severe abdominal distress. What is their digestive system lacking, and what do you recommend be done about it?

Your Turn

Here are some answers. What are the questions?

1. Buffered aspirin _____

2. Skin cancer _____

3. Prostaglandins _____

4. AMP _____

5. pH 7.4 _____

3 Cells: The Basic Units of Life

Active Reading

I. What Are Cells? (57)

1. What are the four most important principles of the cell theory?

   a. _____

   b. _____

   c. _____

   d. _____

2. The outer boundary of the cell is the _____ .

3. The portion of the cell inside the outer boundary and outside the nucleus is the _____. What takes place here? _____. What specialized structures aid this process? _____. What is the fluid portion of this compartment of the cell called? _____.

4. The control center of the cell is the _____. In it lie _____ which contain _____; these direct reproduction, information flow, and _____.

5. The material within the control center is the _____ .

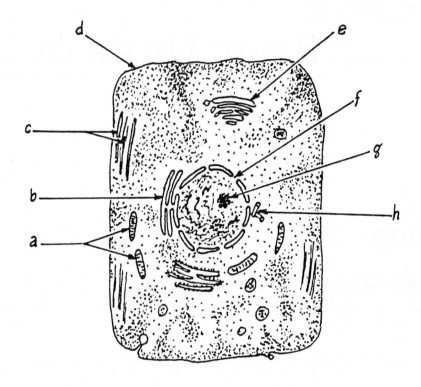

6. The figure above represents a hypothetical cell. Using the letters for reference, identify
   the structures indicated in the figure.

   a. _____

   b. _____

   c. _____

   d. _____

   e. _____

   f. _____

   g. _____

   h. _____

II. Cell Membranes (57)

1. According to the fluid-mosaic model of cell theory, the cell membranes are composed mainly of two types of molecules. What are they? _____

2. Assign the following characteristics to either the <u>head</u> (H) or <u>tail</u> (T) of the phospholipids found in the plasma membrane: (a) polar ( ); (b) contains a glycerol group, a phosphate group, and an alcohol group ( ); (c) hydrophobic ( ); (d) nonpolar ( ); (e) contains fatty acids ( ); (f) hydrophilic ( ); (g) face in toward each other ( ); (h) face out toward the cell exterior ( ).

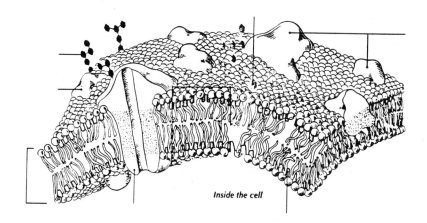

Inside the cell

3. Examine the diagram of a portion of the cell membrane above. Identify and label phospholipids, proteins, glycoproteins, surface carbohydrate, glycolipid, cholesterol.

4. What important role does a cell's specific glycocalyx play? _____

5. How does the presence and concentration of cholesterol affect the molecular bilayer? __

6. List four functions served by the plasma membrane.

    a. _____

    b. _____

    c. _____

    d. _____

7. To which of these functions does the selective permeability of the cell relate? _____

8. Describe the form and function of microvilli. _____

9. Name the four types of cell-adhesion molecules.

    a. _____

    b. _____

    c. _____

    d. _____

10. What role do surface carbohydrates play in the body's defense against bacterial infection? _____

III. Passive Movement Across Membranes (62)

1. When molecules pass through a cell membrane without the use of cellular energy, the movement is called _____

2. Describe the process of simple diffusion. _____

3. What characterizes a state of equilibrium as it pertains to simple diffusion? _____

4. Cite three factors on which the rate of simple diffusion may depend.

   a. _____

   b. _____

   c. _____

5. What is meant by the term <u>concentration gradient</u>? _____

6. How may channels in the plasma membrane facilitate diffusion? _____

7. By what mechanism or mechanisms are closed channels opened? _____

8. Under what conditions is diffusion facilitated? _____

9. Describe an example of a substance being carried across the cell membrane by facilitated diffusion. _____

10. What is osmosis? _____

11. Does osmosis apply to any liquid, or only to water? _____

12. What is osmotic pressure? _____

   The following matching questions apply to the concentration of solutes or of water in the medium <u>surrounding</u> a cell, as compared to solutes or water concentration <u>within</u> the cell.

____ 13. Solute more concentrated in me-   a. isotonic
   dium; therefore the medium is   b. hypertonic
____ 14. Solute less concentrated in medium;   c. hypotonic
   therefore the medium is
____ 15. Solute concentration the same as in
   the cell; therefore the medium is
____ 16. Water concentration less in medium
   than in cell; therefore the medium is
____ 17. Water less concentrated in medium
   than in cell; therefore the medium is

18. Determine in each of the situations described in questions 13–17 whether water would move into the cell (I), out of the cell (O), or neither (N). In question 13 ___, 14 ___, 15 ___, 16 ___, 17 ___

19. What is the driving force in filtration? _____

20. Where in the body does diffusion by filtration occur regularly? _____

IV. Active Movement Across Membranes (65)

1. Does active transport require an input of energy by the cell? _____. If so, where does the energy come from? _____

2. Does active transport work against or with a concentration gradient?_____

3. Describe the events in the active transport process. _____

4. What is endocytosis? _____

5. How is endocytosis achieved?_____

6. What is pinocytosis? _____

7. What is phagocytosis? _____

8. In what form does anything that has been taken into a cell by endocytosis appear? ____

9. Describe receptor-mediated endocytosis._____

10. How is exocytosis achieved? _____

V. Cytoplasm (67)

1. What is cytoplasm? _____

2. What are the two parts or phases of cytoplasm?_____

3. What does cytosol contain?_____

4. What makes up the cytoskeleton? _____

VI. Organelles (67)

1. Complete the table below by naming the major functions of each of the cellular organelles listed:

organelle	major functions
Endoplasmic reticulum	
Ribosomes	
Golgi apparatus	
Lysosomes	
Peroxisomes	
Mitochodria	
Microtubules and microfilaments	
Centrioles	
Cilia and flagella	

VII. The Nucleus (74)

1. Is the nuclear envelope one-layered or two-layered? _____.

2. Name two important aspects of the nucleus.

   a. _____

   b. _____

3. Name a kind of cell that has no nucleus. _____

4. What function do the nuclear pores serve? _____

5. What is the nucleolus composed of, and what is its function? _____

6. Define

    a. nucleoplasm _____

    b. chromatin _____

    c. chromosome _____

VIII.   Cellular Metabolism (77)

1. Name and briefly describe the eight events that lead up to and include the salient elements of metabolism.

    a. _____

    b. _____

    c. _____

    d. _____

    e. _____

    f. _____

    g. _____

    h. _____

IX. The Cell Cycle (77)

1. What is the difference between differentiation and growth? _____

2. What are the phases of the cell cycle? _____

Match the following processes with the figure below.

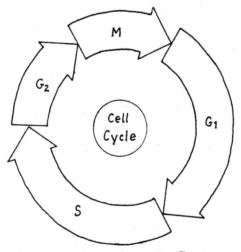

_____ 3. Mitosis

_____ 4. Organelles replicate

_____ 5. Cytokinesis takes place

_____ 6. DNA replicates

_____ 7. Spindle forms

_____ 8. First stage of interphase

_____ 9. Anaphase takes place

a. G_1

b. G_2

c. S

d. M

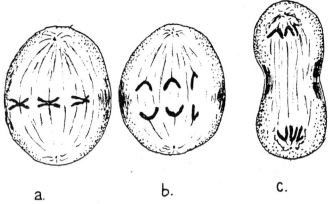

a.   b.   c.

10. Identify the stages of mitosis shown in the figure above.

    a. _____

    b. _____

    c. _____

11. What is the MPF? _____

12. Distinguish between mitosis and cytokinesis. _____

X. How Proteins Are Synthesized (80)

1. Arrange the following events in the order (a – d) in which they occur in protein synthesis: initiation, transcription, termination, elongation

    a. _____

    b. _____

    c. _____

    d. _____

2. What does RNA polymerase do? _____

3. Distinguish between codon, anticodon, initiation codon, and stop codon._____

    _____

    _____

4. What are the roles of (a) mRNA, (b) tRNA, and (c) rRNA?

    a. _____

    b. _____

    c. _____

5. Where is mRNA synthesized? _____

6. Where is mRNA translated? _____

7. What becomes of the polypeptides and/or proteins once they have been synthesized? __

    _____

8. What is the role of DNA in protein synthesis? _____

9. What is the role of DNA in carbohydrate synthesis? _____

10. What is the role of DNA in embryonic development? _____

XI. Cells in Transition: Aging (87)

1. Define the following terms.

   a. gerontology _____

   b. apoptosis _____

   c. necrosis _____

   d. senescence _____

   e. senility _____

   f. free-radical (oxidant) damage _____

   g. cell senescence _____

2. Name four bodily functions that decline with aging.

   a. _____

   b. _____

   c. _____

   d. _____

XII. Cells Out of Control: Cancer (90)

1. Distinguish between benign and malignant neoplasms. _____

2. Define or briefly describe

   a. carcinoma _____

   b. sarcoma _____

   c. mixed-tissue neoplasm _____

   d. leukemia _____

   e. metastasis _____

   f. dysplasia _____

   g. carcinogens _____

   h. oncogene _____

3. What are the genes p53 and Rb? _____

4. What are the traditional methods of treating cancer? _____

5. List the seven most common sites for cancer.

   a. _____

   b. _____

   c. _____

   d. _____

   e. _____

   f. _____

   g. _____

XIII. When Things Go Wrong (96)

1. What is progeria? How do adult and juvenile progeria differ? _____

_____

Key Terms

active transport 65
cancer 90
cell cycle 77
cell theory 57
centriole 73
chromosome 76
cilia 73
cytokinesis 80
cytoplasm 57, 67
cytoskeleton 67, 72
differentiation 78
diffusion 62
endocytosis 65
endoplasmic reticulum 69

exocytosis 65
facilitated diffusion 64
flagella 73
fluid-mosaic model 57
Golgi apparatus 69
lysosome 70
messenger RNA 86
metabolism 77
microfilament 73
microtubule 72
mitochondrion 71
mitosis 80
neoplasm 90
nuclear envelope 75

nucleolus 75
nucleoplasm 57
nucleus 57, 74
organelle 67
osmosis 64
passive movement 62
phagocytosis 65
plasma membrane 57
protein synthesis 80
replication 78
ribosome 69
semipermeability 61

Post Test

Multiple Choice

____ 1. The basic principles of the cell
        theory were formulated by

a. Hardy and Weinberg
b. Hamilton and Boyd
c. Charles Darwin
d. Watson and Crick
e. Schleiden and Schwann

____ 2. The fluid portion of a cell's interior
        is the

a. nucleoplasm
b. karyoplasm
c. cytosol
d. cytoplasm
e. plasmasol

____ 3. According to the fluid-mosaic model
        of membrane structure, the main
        framework of the plasma membrane
        is a

a. hydrophobic mesh
b. protein/lipid mosaic
c. hydrophilic monolayer
d. phospholipid bilayer
e. carbohydrate film

___ 4. The surface glycolipids and glycoproteins of the plasma membrane
may serve as recognition or receptors sites for

a. nerve dendrites
b. ATP
c. hormones
d. hydrolytic enzymes
e. plasma cells

___ 5. The flexibility of the plasma membrane is determined by the ratio of

a. cholesterol to phospholipids
b. cholesterol to glycoproteins
c. phospholipids to carbohydrates
d. glycoproteins to glycolipids
e. none of these

___ 6. Among the vital functions of the
plasma membrane is

a. keeping the inside and outside of the cell
separate
b. separating cells from one another
c. providing a surface for chemical reactions
to occur on
d. regulating the flow of materials into and
out of cells
e. all of the above

Identification

7. CAMs _____

8. concentration gradient _____

9. carrier protein _____

10. osmosis _____

11. hypertonic _____

12. energy source for active transport _____

13. pinocytosis _____

14. ER _____

15. cytoskeleton _____

Matching

___ 16. Major site of ATP synthesis
___ 17. Small membrane-bound sacs containing acid hydrolases
___ 18. Major site of ribosome preassembly.
___ 19. RNA/protein particles used in
protein synthesis
___ 20. Microtubular bundles that function
in cell division
___ 21. Cytoplasmic, saclike organelle that
functions in the "packaging" and
secretion of glycoproteins

a. ribosomes
b. Golgi apparatus
c. lysosomes
d. mitochondria
e. centrioles
f. nucleolus
___ 22. Cell division

_____ 23. First step in actual protein synthesis
_____ 24. Preliminary process in protein synthesis
_____ 25. G_1, S, G_2, M
_____ 26. Has A and P sites

a. cell cycle
b. mitosis
c. cytokinesis
d. transcription
e. translation
f. ribosome

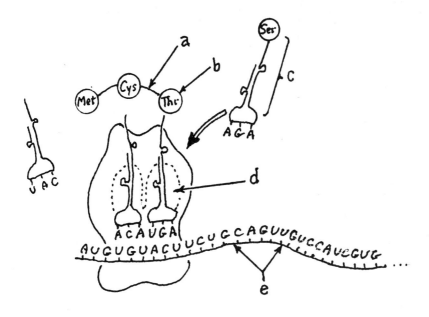

27. Identify the structures indicated in the figure above.

a. _____

b. _____

c. _____

d. _____

e. _____

Multiple Choice

_____ 28. The death of cells as a result of the action of particular "cell-death genes" is known as

a. senility
b. necrosis
c. senescence
d. apoptosis
e. senile dementia

_____ 29. Aging is believed to result from

a. the action of free radicals
b. cellular inability to divide indefinitely
c. apoptosis
d. insufficient activity of growth factors
e. all of the above

Complete each sentence below with its appropriate term

_____ 30. Environmental agents that cause cellular changes leading to cancer are known as

_____ 31. A gene that normally regulates a cellular process, after mutation, can transform a cell into a cancer cell is termed a(n)

_____ 32. A synonym for tumor is

_____ 33. A neoplasm that is encapsulated is called

_____ 34. An unencapsulated, spreading neoplasm is said to be

_____ 35. A cancer of connective tissue or muscle tissue is called a(n)

_____ 36. A _____ is a cancer that originates in epithelial tissue.

_____ 37. The word that means the spread of cancer cells is

_____ 38. After mutation from its original state, a(n) _____ is capable of transforming a normal cell into a cancer cell.

a. benign
b. malignant
c. neoplasm
d. oncogene
e. carcinoma
f. sarcoma
g. metastasis
h. carcinogen
i. proto-oncogene

Integrative Thinking

1. Recent studies of African pigmies have revealed that they are not deficient in growth hormone, as previously thought. Normal amounts are found circulating in their bloodstream. Assuming that the anomaly is cellular, rather than systemic, where should the investigators look next? Explain.

2. A young woman develops a condition characterized by fatigue that grows progressively worse. After standard examinations reveal nothing wrong in any of the systems – nervous, muscular, endocrine – the physicians begin to suspect that the problem lies at the cellular level. Where in the cells would one look? Explain.

3. The cells listed below differ from each other in the degree of development of certain organelles. Which organelle would one expect to be especially prominent in terms of number or relative size in each of these?

a. Pancreatic cell that secretes digestive enzymes _____

b. A skeletal muscle cell that uses lots of energy _____

c. An epithelial cell of the intestine that specializes in absorption _____

d. A cell of the immune system that engulfs bacteria and/or tissue debris _____

e. A dying cell _____

Your Turn

Here is a new twist: Make up two multiple-choice questions. The first one will be the usual kind, with only one correct answer. The other one will ask for the only <u>wrong</u> answer. Doubtless you have encountered this type before. However, they are a little harder to make up, but you learn a lot from them. The reason is that there is an infinite number of wrong answers and not many right ones. Bear in mind: The questions must be based on material in this chapter, and each question must have five choices. The best questions submitted might be seen again in the next test!

4 Tissues of the Body

Active Reading

I. Development of Primary Germ Layers (101)

1. What is meant by the term *differentiation?* _____

2. Using the table, identify the three primary germ layers and the major organs that will develop from them.

Layer	Major Organs or Systems

3. Identify the structures indicated in the figure below.

   a. _____

   b. _____

   c. _____

   d. _____

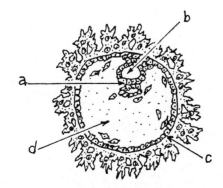

II. Epithelial Tissues: Form and Function (101)

1. What are the six typical functions of epithelial tissues?

    a. _____

    b. _____

    c. _____

    d. _____

    e. _____

    f. _____

2. What are junctional complexes? _____

Match the following. (Recall that with either matching or multiple-choice questions, a given answer may be used only once, more than once, or not at all.)

____ 3. A homogeneous extracellular material composed mainly of glycoproteins

____ 4. A kind of junction between adjacent cell membranes that consists of a crisscrossed network of intercellular filaments; common in skin

____ 5. Intimate and close connection between cells involving little or no extracellular space between them

____ 6. Formed from several links of channel protein that connect the plasma membranes of adjoining cells

____ 7. A network of extracellular fibers and a homogeneous protein material

a. tight junctions
b. spot desmosomes
c. gap junctions
d. ground substance
e. matrix

8. Label the figure below, using the letters for reference.

    a. _____

    b. _____

    c. _____

    d. _____

    e. _____

    f. _____

    g. _____

    h. _____

    i. _____

Match the following.

___ 9. The shape of surface cells of most epithelia

___ 10. A component of the membrane on which epithelium rests

___ 11. Brush border

___ 12. Intracellular strands that strengthen the inner framework of certain epithelial cells

___ 13. A component of basement membranes

___ 14. A rounded, hollow shape

a. basal lamina
b. tonofilaments
c. microvilli
d. squamous
e. alveolar

15. What is the most likely function of epithelia that have microvilli on their free surface? _

_____

16. Is epithelial tissue, generally speaking, well vascularized (vascular = having blood vessels)? _____

17. Do epithelia ever contain nerve fibers? _____

18. Where is one likely to find cilia? _____

19. Describe the basic structural feature of each of the following classes of epithelia.

   a. simple _____

   b. stratified _____

   c. squamous _____

   d. cuboidal _____

   e. columnar _____

20. Give one example and the general function of each of the following classes of epithelia.

   a. simple squamous _____

   b. simple cuboidal _____

   c. simple columnar _____

   d. stratified squamous _____

   e. stratified cuboidal _____

   f. stratified columnar _____

   g. pseudostratified _____

   h. transitional _____

21. List three functions of basement membranes.

   a. _____

   b. _____

   c. _____

22. Complete the table below by naming in the first column the types of glands, in the third column the shape of the secretory portion of each type of gland, and in the fourth column the location of each type of gland.

TABLE 4.1 CLASSIFICATION OF MULTICELLULAR EXOCRINE GLANDS

Type of gland	General shape	Shape of secretory portion	Location

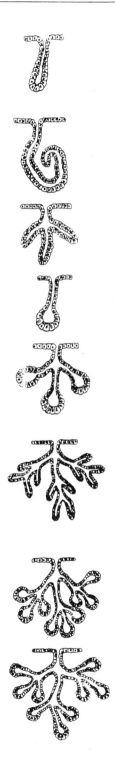

23. Define the following terms.

    a. exocrine _____

    b. endocrine _____

    c. merocrine _____

    d. holocrine _____

    e. mucous gland _____

    f. serous gland _____

    g. mixed gland _____

    h. goblet cell _____

III. Connective Tissues (112)

1. Do connective tissues have all these functions: (a) support other cells and organs, (b) form protective sheaths around hollow organs, (c) storage, (d) repair, (e) transport? Y/N_

2. Besides connective tissue proper, what other classes ("improper"?) are there? _____
_____

3. What are the three types of connective tissue proper?

    a. _____

    b. _____

    c. _____

4. Place a check mark before items that are components of connective tissues.

    __ fibers

    __ glands

    __ ground substance

    __ cells

    __ some extracellular fluid

5. Name two diagnostic features of collagenous fibers.

    a. _____

    b. _____

6. Name two diagnostic features of reticular fibers.

    a. _____

    b. _____

7. Name two diagnostic features of elastic fibers.

    a. _____

    b. _____

8. What are the functions of each of the following?

    a. fibroblasts _____

    b. adipose cells _____

    c. macrophages _____

    d. reticular cells _____

    e. plasma cells _____

    f. mast cells _____

    g. leukocytes _____

9. Which of the cells named in question 8 are fixed cells? _____

_____

10. Which of the fixed cells are sometimes wandering cells? _____

11. Which of the cells listed are wandering cells? _____

12. Which cells can undergo diapedesis? _____

13. Where are plasma cells usually found? _____

14. What are the secretory granules that are found in mast cells, and what are their functions?

_____

15. What is the distinction between <u>areolar</u> (loose) connective tissue and <u>collagenous</u> (dense) connective tissue? _____

16. Place a check mark before each of the items that characterizes cartilage.

    __ chondrocytes

    __ unilocular fat cells

    __ multilocular fat cells

    __ lacunae

    __ ground substance

    __ fibers

17. How does bone differ from cartilage? _____

18. In what ways are bone and cartilage similar? _____

19. Why is blood classified as a kind of connective tissue? _____

IV. Muscle Tissue: An Introduction (121)

1. Compare skeletal muscle and cardiac muscle. _____

2. Why are they considered to be tissues? _____

3. Compare skeletal muscle and smooth muscle. _____

V. Nervous Tissue: An Introduction (121)

1. What is the essential function of nervous tissue? _____

2. Compare the functions of neurons, neuroglia, and peripheral glial cells. _____

_____

VI. Membranes (121)

1. Name the three kinds of membranes and where each occurs.

   a. _____

   b. _____

   c. _____

2. Are membranes epithelial tissue or connective tissue? _____

3. What are the functions of mucous membranes? _____

4. What is mesothelium? _____

Match the following.

____ 5. Consists of water, salts, and glyco-
    proteins
____ 6. The layer that lines the abdominal
    wall
____ 7. Covers the lungs
____ 8. Covers the small intestine
____ 9. Lines the thoracic cavity
____ 10. Contains the heart

    a. pericardium
    b. visceral peritoneum
    c. parietal peritoneum
    d. mucus
    e. pleura

11. Fill in the blank spaces in the following running description: "Organs of the abdominopelvic cavity are suspended from the cavity wall by fused layers of visceral peritoneum that are called _____, one of which, the _____ suspends the stomach. Others suspend the spleen and portions of the small intestine. The one that suspends the uterus is known as the _____. The kidneys and pancreas, although their location is _____, are also supported by _____, which attach them to the posterior wall of the abdominopelvic cavity."

12. What is the name of the membranes that line the cavities of movable joints, such as the knee and elbow? _____. What are these membranes composed of? ____

13. Name and describe the fluid that is associated with these membranes. _____
_____

VII. When Things Go Wrong (125)

1. Define pathogenesis. _____

2. What is the etiology (cause) of scurvy? _____

3. What is rheumatoid arthritis? How is it caused? _____

4. What is the etiology of arteriosclerosis? _____

5. Distinguish between needle biopsy, aspiration biopsy, endoscopic biopsy, open biopsy, and excisional biopsy. _____
_____

6. What are the symptoms of spider fingers? _____

7. What is the technical name for spider fingers? _____

8. Describe the course of systemic lupus erythromatosis. _____

9. What characterizes edema? _____

10. Describe the etiology and course of gangrene. _____

   _____

11. What is meant by tissue grafting? Name and distinguish between the four basic types of tissue transplantations.

   a. _____

   b. _____

   c. _____

   d. _____

Key Terms

adipose tissue 115	epithelial tissue 101	microvilli 104
basement membrane 104	exocrine gland 105	mucous membrane 121
cartilage 116	fixed cells 115	primary germ layers 101
chondocyte 116	ground substance 114	reticular tissue 114
connective tissue 112, 116	junctional complexes 103	serous membrane 123
dense connective tissue 116	loose connective tissue 116	simple epithelial tissue 104, 106
differentiation 101	macrophage 115	stratified epithelium 104, 108
ectoderm 101	matrix 112	synovial membrane 124
elastic tissue 114	membrane 121	tissues 101
endoderm 101	mesoderm 101	wandering cells 115

Post Test

Matching

____ 1. Nervous system

____ 2. Blood

____ 3. Outer layers of skin

____ 4. Lining of stomach

____ 5. Bone

____ 6. Lining of lungs

____ 7. Muscle

a. ectoderm
b. endoderm
c. mesoderm

8. Identify the six classes of epithelia drawn below, using the chart for your answer. Identify the seventh class, which is not illustrated, and name a typical location in the body for each.

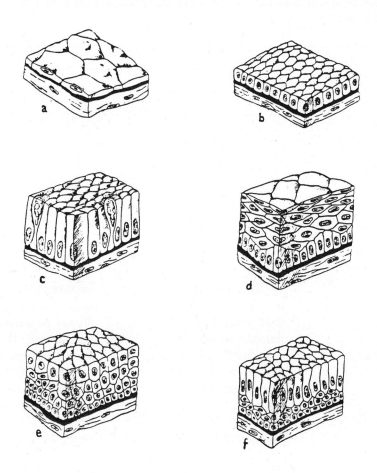

	Class of Epithelium	Typical Location
a.		
b.		
c.		
d.		
e.		
f.		
g.		

Multiple Choice

____ 9. A group of cells of similar structure and serving the same general or identical function defines a(n)

    a. organ
    b. connective tissue
    c. epithelium
    d. system
    e. tissue

____ 10. A tissue that forms sheets that cover body surfaces or line body cavities or ducts is

    a. skin
    b. connective tissue
    c. epithelium
    d. membrane tissue
    e. basement membrane

____ 11. A homogeneous, extracellular material that is composed mainly of glycoproteins is known as

    a. collagen
    b. elastin
    c. intercellular matrix
    d. serous membrane
    e. ground substance

____ 12. The plasma membranes of adjoining epithelial cells are often connected to each other by several links of channel proteins. This kind of connection constitutes a(n)

    a. gap junction
    b. spot junction
    e. basal lamina
    d. tight junction
    e. spot desmosome

____ 13. A network of extracellular reticular fibers plus a sheet of homogeneous protein composition that together provide support for a layer of epithelial cells is called a(n)

    a. basal lamina
    b. tight junction
    c. brush border
    d. ground substance
    e. galactoprotein

____ 14. Intracellular fiberlike strands that provide support for cellular structure at spot desmosomes are known as

    a. reticular fibers
    b. tonofilaments
    c. matrix
    d. channel proteins
    e. Golgi apparatus

____ 15. The blood supply for epithelia is provided by

    a. a penetrating capillary network
    b. diffusion of nutrients and metabolites from capillaries in the underlying connective tissue
    c. a network of fine nervous tissue elements
    d. a porous system of epithelial blood vessels
    e. none of the above

____ 16. In addition to its role in providing support and protection to various anatomical structures, epithelium also is especially noteworthy for its _____ functions.

    a. glandular
    b. vascular
    c. immunological
    d. sensory
    e. both a and d

____ 17. Goblet cells are a prominant feature in epithelia of the

    a. intestinal tract
    b. respiratory tract
    c. stomach
    d. conjunctiva of the eye
    e. all of the above

46   Chapter 4

Matching

___ 18. Glands that secrete a watery sub-
       stance generally containing enzymes
___ 19. Glands that release their secretion
       without breaking the plasma mem-
       brane
___ 20. Secretion consists of whole cells
___ 21. The pancreas as an exocrine gland
___ 22. Glands that produce a thick, lubri-
       cating, glycoprotein secretion

a. mucous
b. holocrine
c. merocrine
d. serous
e. mixed

___ 23. Yellow, branching fibers
___ 24. Thin, delicate fibers
___ 25. Fibers that form thick bundles
___ 26. Homogeneous material composed of
       glycoproteins and proteoglycans
___ 27. Principal fibers of tendons and
       ligaments

a. collagenous
b. ground substance
c. elastic
d. reticular
e. matrix

___ 28. Cells that engulf and destroy foreign
       bodies in bloodstream and tissues
___ 29. Main producers of antibodies
___ 30. Synthesize and store lipids
___ 31. Synthesize matrix materials and
       assist wound healing
___ 32. Produce heparin and histamine

a. fibroblasts
b. adipose cells
c. macrophages
d. mast cells
e. plasma cells

___ 33. A rather fluid connective tissue with
       irregularly arranged fibers
___ 34. In intervertebral disk and knees
___ 35. Pearly blue-white; most common
       type of cartilage
___ 36. Bone
___ 37. Epiglottis, external ear, larynx

a. areolar
b. hyaline
c. vascularized
d. fibrocartilage
e. elastic cartilage

___ 38. The membrane that lines the ab-
       dominal wall
___ 39. Lines the thoracic cavity
___ 40. Contains the heart
___ 41. Continuous with mesenteries
___ 42. Lines cavities of movable joints

a. pericardium
b. visceral peritoneum
c. parietal peritoneum
d. synovial membrane
e. pleura

Complete the following account by selecting the appropriate term from the list.

Collagen is believed to be the most abundant protein in the world. Not surprisingly, when something goes wrong with collagen, the results are dire. A few rather common cases are these: If collagen fibers are not formed, owing to vitamin C deficiency, ____ will result. When collagen in the cartilage of a synovial joint is destroyed by autoimmune disease or other causes, painful ____ is the consequence. The vascular disease ___ is exacerbated by the secretion of large amounts of collagen by the smooth muscle cells that surround the larger blood vessels.

a. rheumatoid arthritis
b. scurvy
c. arteriosclerosis
d. cystic fibrosis

Integrative Thinking

1. Two very different kinds of nonvacularized tissue are described in this chapter: epithelium and cartilage. Why do you think they do not contain blood vessels?

2. The brain possesses a unique condition known as the blood-brain barrier, which is exceedingly selective as to what substances can be transported across it. Glucose, oxygen, and carbon dioxide can diffuse across it very readily, but little else can. Based on what you have learned in this chapter, suggest at least two anatomical explanations for this barrier.

Your Turn

Here are some answers. What are the questions?

1. Holocrine _____

2. Thermography _____

3. Brown fat _____

4. Gas gangrene _____

5. Mesothelium _____

5 The Integumentary System

Active Reading

I. Skin (133)

1. What is the skin's approximate area? _____. What is the skin's approximate thickness?_____.

2. What three main parts make up the skin?

   a. _____

   b. _____

   c. _____

3. The outer layer of skin, or _____, is composed of what type of tissue (check your choice)?

   a. simple cuboidal epithelium

   b. goblet cells

   c. stratified squamous epithelium

   d. collagenous connective tissue

4. Label the five strata of epidermal tissue in the figure below.

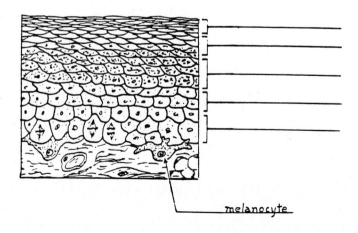

melanocyte

5. Which two strata are collectively known as the *stratum germinativum*? _____. Why are they called this? _____. What three strata can be found covering most areas of the body? _____

6. Describe the role of keratin in the epidermis. _____

7. What is the dermis? _____

8. Name the two layers of the dermis.

    a. _____

    b. _____

How do they differ? _____

9. Identify the following with either the papillary layer of the dermis (P), the reticular layer of the dermis (R), or both layers of the dermis (PD): loose connective tissue ___, dense connective tissue ___, collagenous fibers ___, nerve endings ___, fingerprint ridges ___, cleavage lines ___, elastic fibers ___, tiny, fingerlike projections into the epidermis ___.

10. What is the hypodermis? _____

11. By what other name is the hypodermis called? _____

12. What sensory receptors are found in the skin? _____

13. Describe five major functions of the skin.

    a. _____

    b. _____

    c. _____

    d. _____

    e. _____

14. What three agents give skin its color?

    a. _____

    b. _____

    c. _____

15. In an albino, one might expect the following to be dysfunctional. _____

16. How does prolonged exposure to sunlight affect the skin? _____

17. List the four dermal sources of blood for the skin.

    a. _____

    b. _____

    c. _____

    d. _____

18. Describe the four-step process for the healing of a wound. Include in your description the following terms: *platelets, fibrinogen, fibroblasts, neutrophils, monocytes,* and *scab.*

    a. _____

    b. _____

    c. _____

    d. _____

19. Identify the terms FGF, TGFb, and PDGF. What role does each play in wound healing?

    _____

    _____

20. Histamine, lysosome enzymes, edema, and dead white blood cells (pus) contribute to the _____, causing the symptoms of redness, swelling, and pain.

II. Glands of the Skin (141)

1. What glands produce sweat? _____

2. Where on the body are eccrine glands located? _____

3. In what layers of the skin are these glands located? _____

4. What function does perspiration serve? _____

5. How are apocrine glands and eccrine glands stimulated?

    a. _____

    b. _____

6. Odiferous glands in the armpits, around the nipples, and in anal and genital regions are _

    _____ .

7. What glands secrete earwax? _____

8. Label the figure below as indicated.

    a. _____

    b. _____

    c. _____

    d. _____

    e. _____

    f. _____

___ 9. What does sebum function as?

    a. a pheromone
    b. an emollient
    c. a permeability barrier and protective agent
    d. a nuisance for many self-conscious teenagers
    e. all of the above

10. What is a blackhead?

11. What is a whitehead?

III. Hair (143)

1. From what layer of skin does hair develop?_____

2. Name and briefly describe the three layers of a typical hair shaft.

    a. _____

    b. _____

    c. _____

3. In the figure below, identify the following structures: hair shaft, root, follicle, bulb, papilla, matrix, sebaceous gland, arrector pili muscle. Indicate whether each structure is epidermal or dermal.

___ 4. Curly hair has follicle openings that are

    a. oval
    b. spiral-shaped
    c. round

5. Identify the following statements as either true (T) or false (F).

    a. All the body's hair follows the same growth cycle____

    b. Hair grows faster at night than during the day____

    c. Causes of baldness are genetic, and may include high amounts of testosterone and low amounts of estrogen____

    d. Once a follicle sheds its hair, that follicle becomes inactive and closes up____

    e. Hair, like nails, derives from the same embryonic tissue as the skin____

IV. Nails (145)

1. Identify the free edge, nail bed, matrix, and eponychium on the figure below.

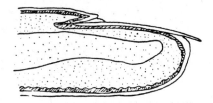

2. What protein accounts for the hardness and strength of the nails? _____

3. What are the functions of nails? _____

4. Are nails formed from epidermal or dermal cells? _____

5. Summarize the structures described thus far in the following table.

Structure	Epidermal/Dermal	Function
Skin		
Glands a. eccrine b. apocrine c. sebaceous		
Hair		
Nails		

V. Effects of Aging on the Integumentary System (145)

1. Define the following terms.

    a. lentigines _____

    b. cherry angiomas _____

    c. seborrheic keratoses _____

2. List four other effects of aging on the skin.

    a. _____

    b. _____

    c. _____

    d. _____

3. List three effects of aging on the hair.

    a. _____

    b. _____

    c. _____

VI. Developmental Anatomy of the Integumentary System (146)

1. From what embryonic germ layer are the following derived?

    a. the epidermis _____

    b. the dermis _____

2. What is the periderm? _____

3. From what embryonic tissue and fibers do the dermal papillae develop? _____

4. What is the vernix caseosa and from what is it formed? _____

5. What is the function of the vernix caseosa? _____

6. What are (a) lanugo and (b) vellus?

    a. _____

    b. _____

7. What is terminal hair? _____

VII. When Things Go Wrong (147)

____ 1. The Lund-Browder method and the rule of nines are methods of estimating
    a. the extent of burns on the body
    b. the depth of burns in the skin and underlying tissue
    c. both of the above

____ 2. The following are causes of burns
    a. chemicals
    b. electricity
    c. radioactivity
    d. heat
    e. all of the above

3. Describe the depth of tissue damage and major effects of

    a. first-degree burns _____

    b. second-degree burns _____

    c. third-degree burns _____

4. Outline the treatment recommended for burns:

    a. first-degree burns _____

    b. second-degree burns _____

    c. third-degree burns _____

5. Identify the secondary problems that are consequences of burns. _____

_____

6. Describe the symptoms and causes for the following skin disorders.

    a. acne vulgaris _____

    b. bedsores _____

    c. birthmarks and moles _____

    d. psoriasis _____

    e. poison ivy, poison oak, or poison sumac _____

    f. warts _____

7. What are the two most common forms of skin cancer?

    a. _____

    b. _____

8. What is the most serious type of skin cancer? _____

9. Name the common causes of skin cancer. _____

10. What does this describe: "This type of skin cancer starts as a small, dark growth, gradually becomes larger, develops an irregular outline, changes color, becomes ulcerated, and bleeds easily"? _____

Key Terms

apocrine gland 141	epidermis 133	keratin 133 f.
cleavage lines 136	follicle 144	nails 145
dermis 133	hair 143	sebaceous gland 141
eccrine gland 141	hypodermis 136	sudoriferous gland 141

Post Test

Multiple Choice

____ 1. The integumentary system consists of the

    a. epidermis and endodermis
    b. epidermis and dermis
    c. epidermis, dermis, and hypodermis
    d. epithelium and dermis
    e. epithelium, dermis, and hypodermis

____ 2. The epidermis consists of the strata corneum, lucidum, granulosum, spinosum, and

    a. germinativum
    b. basale
    c. papillium
    d. reticulum
    e. none of these

____ 3. The skin covering most of the body, except the palm of the hand and sole of the foot, consists only of the
   a. stratum corneum
   b. strata corneum and lucidum
   c. strata corneum, lucidum, and granulosum
   d. strata corneum and germinativum
   e. strata corneum and basale

____ 4. The stratum corneum owes most of its physical properties to
   a. cholesterol
   b. saturated fatty acids
   c. a variety of lipids
   d. soft keratin
   e. hard keratin

____ 5. The thickest layer of the skin generally is the
   a. stratum germinativum
   b. epidermis
   c. dermis
   d. hypodermis
   e. papillary layer

____ 6. The layer of the dermis that includes capillary loops, tactile corpuscles of Meissner, and are responsible for the ridge patterns of fingerprints
   a. reticular layer
   b. papillary layer
   c. cleavage (Langer's) layer
   d. hypodermis
   e. adipose layer

____ 7. The layer in which are embedded many blood and lymphatic vessels, nerve endings, fat cells, sebaceous glands, hair roots, and smooth muscle fibers
   a. reticular layer
   b. papillary layer
   c. hypodermal layer
   d. adipose layer
   e. both c and d

____ 8. Among the many functions of the skin are
   a. temperature regulation
   b. excretion
   c. synthesis of vitamin D
   d. sensory reception
   e. all of these

____ 9. The dermis derives embryologically mainly from
   a. surface ectoderm
   b. parietal ectoderm
   c. mesodermal somites
   d. mesenchyme
   e. endoderm

____ 10. The color of skin is due principally to
   a. melanin
   b. carotene
   c. both melanin and carotene
   d. environmental soot
   e. liver spots

Matching

____ 11. Provide nourishment to the epidermis
____ 12. Proteins that promote wound healing
____ 13. Subcutaneous touch receptors
____ 14. An apocrine secretion
____ 15. Earwax
____ 16. A significant factor in wrinkling of
the skin
____ 17. Located in the papillary layer

a. Meissner's corpuscles
b. elastin → elacin
c. capillary network
d. PDGF, FGF, TGFb
e. cerumen

____ 18. Associated with sebum secretion
____ 19. Responsive to heat
____ 20. Ceruminous glands are of this kind
____ 21. Also known as odoriferous gland
____ 22. The type of gland that is responsive
to stress
____ 23. Responsible for so-called body odor
____ 24. Includes mammary glands
____ 25. The type that includes sweat glands

a. eccrine gland
b. apocrine gland
c. sebaceous gland
d. mucous gland

____ 26. The central core of a hair
____ 27. A coating of squamous cells
____ 28. The portion of hair that protrudes
from the skin
____ 29. The portion that is embedded in the
skin
____ 30. The thickest layer of the hair, and
the part that includes that hair's
pigment

a. shaft
b. root
c. medulla
d. cortex
e. cuticle

____ 31. Consists of three sheaths: inner,
outer, and connective tissue
____ 32. Composed of smooth muscle
____ 33. Fetal hair
____ 34. The expanded, inner tip of the hair
____ 35. The epithelial cells that comprise the
preceding (#34)

a. bulb
b. matrix
c. follicle
d. arrector pili
e. lanugo

36. Identify three somewhat common causes of premature hair loss.

a. _____

b. _____

c. _____

37. What function(s) does scalp hair serve? _____

38. What is the lunula, and why is it white? _____

39. What is the principal chemical component of fingernails? _____

40. What does the cuticle represent? _____

41. What is the common name of each of the following?

   a. lentigines _____

   b. open comedones _____

   c. acne vulgaris _____

   d. decubitus ulcers _____

   e. nevus _____

Multiple Choice

___ 42. A condition of the integument that is marked by red, dry, and elevated lesions that are covered with silvery, scaly patches

   a. acne
   b. psoriasis
   c. melanoma
   d. warts
   e. squamous cell carcinoma

___ 43. Malignant growths of epithelial origin are classified as

   a. malignant melanomas
   b. squamous cell carcinomas
   c. carcinomas
   d. sarcomas
   e. basal cell epitheliomas

___ 44. The two most common forms of skin cancer are

   a. squamous cell and basal cell carcinomas
   b. alpha cell and beta cell epitheliomas
   c. melanomas and sarcomas
   d. carcinomas and sarcomas
   e. none of the above

___ 45. An example of benign epithelial tumors

   a. basal cell carcinoma
   b. squamous cell carcinoma
   c. psoriasis
   d. malignant melanoma
   e. warts

___ 46. Doubtless the most dangerous of skin lesions

   a. malignant melanoma
   b. squamous cell carcinoma
   c. basal cell carcinoma
   d. psoriasis
   e. warts

Integrative Thinking

1. Perhaps the most excruciating kind of injury is a second-degree burn. Third-degree burns are more life-threatening but are not as painful. How do you account for the difference?

2. It used to be customary to apply butter or greasy emollients to burns, especially to first-degree burns. This treatment is no longer recommended. In fact, it is vehemently discouraged. Why?

3. When the skin is cut, say, with a knife, it repairs itself. Lost or damaged cells are replaced by division of cells adjacent to the wound. The same sort of response follows most second-degree burns, but such is not the case for third-degree burns of no greater area. How do you account for the difference? What special treatment do third-degree burns require, and why?

Your Turn

Make up questions for which the following could be the answers.

1. Because they have more carotene and less melanin.

2. The melanin is more densely distributed and the melanocytes are more productive.

3. The hair just looks longer because the skin shrinks.

4. Because there is no tyrosinase.

5. Because the baby would get hair caught between its teeth(if it has any teeth!)

   Now it's really your turn. Make up some more answers and questions!

6 Bones and Bone Tissue

Active Reading

I.   Types of Bones and Their Mechanical Functions (157)

1. What are the two main types of supportive connective tissue?

    a. _____

    b. _____

2. What is another name for bone tissue? _____

3. In addition to providing support and protection, bone tissue serves as a storehouse for
    _____ and _____. Bone marrow is the site for the manufacture of
    _____ and _____ cells.

4. List the three mechanical aspects of bone function.

    a. _____

    b. _____

    c. _____

5. Classify the following bones as either long, short, flat, irregular, sesamoid, or accessory.
    Where possible, name the specific bone described.

    a. Embedded within the tendon of the quadriceps femoris, this bone protects the
        tendon, helps it overcome compression forces, and increases the mechanical
        efficiency of the knee joint. _____

    b. This bone helps to form a protective shell for the brain. _____

    c. While this type of bone is most commonly found in the feet, it also occurs as small
        bony clusters between the joints of the flat bones of the skull. _____

    d. This bone, the longest bone in the body, acts as a lever when pulled by contracting
        leg muscles. _____

    e. A bone that is shaped to allow passage of the spinal cord and has extensions from
        its main bony elements that serve as sites for muscle attachments. _____

    f. These bones, of which the carpal bones and tarsal bones are the only examples,
        are irregularly shaped bones of roughly equal length, width, and thickness. _____

6. Mark the following statements as either true or false:

 a. Short bones occur where only limited movement is required ____

 b. Accessory bones are formed when developing bones fuse together ____

 c. Long bones, acting as levers, make it possible for the body to move ____

 d. Flat bones are so-called because they sometimes resemble flat seeds ____

 e. An example of bones that were named after flat seeds is the patella ____

II. Gross Anatomy of a Typical Long Bone (159)

1. On the figure below, label the proximal and distal epiphyses, the metaphysis, the diaphysis, the epiphyseal plate, and the articular cartilage.

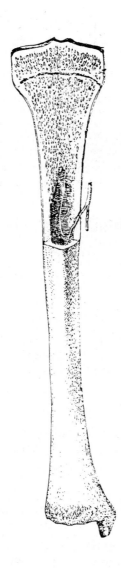

2. After birth, where does all growth in long bones occur? _____

Match the following pairs.

___ 3. Medullary cavity	a.	myeloid tissue
___ 4. Trabeculae	b.	contains yellow marrow
___ 5. Red marrow	c.	hyaline cartilage
___ 6. Epiphyseal plate	d.	opening in compact bone
___ 7. Nutrient foramen	e.	spongy bone

8. Assign the following features to either the endosteum or the periosteum.

    a. covers the outer surface of the bone _____

    b. lines the medullary cavity _____

    c. fibrous membrane that aids in fracture healing _____

    d. attached to bone by collagenous periosteal perforating fibers _____

    e. covers the trabeculae of spongy bone tissue _____

    f. contains nerves, lymphatic vessels, and capillaries _____

III. The Histology of Bone (Osseous) Tissue (160)

1. Unlike the human body as a whole, bones are only ___ percent water.

2. Name the two inorganic salts, found within the matrix of ground substances, that are primarily responsible for the rigidity of bone.

    a. _____

    b. _____

3. These salts form _____crystals, which ionize when required to provide the amounts of calcium and/or _____ the body needs.

4. What is the layer of spongy tissue between the tables of the flat bones of the cranium called? _____

Match the following pairs.

___ 5. Perforating canal	a.	osteocyte
___ 6. Central canal	b.	Volkmann's canal
___ 7. Compact bone cell	c.	Sharpey's fibers
___ 8. Periosteal perforating fibers	d.	osteon
___ 9. Haversian system	e.	Haversian canal

___10. Spongy (cancellous) bone tissue gains its strength from the arrangement of which of the following? (a) lacunae, (b) lamellae, (c) alumnae, (d) trabeculae. (Caesar would have loved to see all these ae's!)

11. Identify the following on the figures below: lamellae, central canal, lacuna, canaliculi, osteocyte, trabeculae.

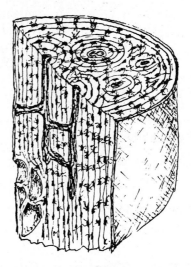

IV. Bone Cells (163)

1. Describe the structure and function of the following cells.

   a. osteocytes _____

   b. osteoblasts _____

   c. osteoclasts _____

   d. bone-lining cells _____

   e. osteogenic (osteoprogenitor) cells _____

2. What is osteoid, and from where does it come? _____

VI. The Physiology of Bone Formation: Ossification (164)

1. What are the two ways bones develop in the human embryo?

   a. _____

   b. _____

2. Define the following.

   a. mesenchyme _____

   b. trabeculae _____

   c. ossification center _____

3. Describe the role of osteoblasts in intramembranous ossification. _____

   _____

___ 4. Which of the following form through the process of intramembranous ossification? (a) skull, (b) scapula, (c) femur, (d) humerus

5. True or false? During the process of endochondral ossification, the hyaline cartilage becomes transformed into bone tissue. ___

6. Describe the process of endochondral ossification at the cellular level. Include in your description the terms chondrocytes, alkaline phosphatase, cartilaginous spicules, osteoid, osteoblast, periosteal bud, pluripotent cell, osteoclast, and osteocyte. _____

_____

_____

7. In what area of long bones does endochondral ossification begin? _____

VI. The Physiology of Bone Growth (169)

1. On the figures below, identify the following: cartilage, periosteum, perichondrium, primary center of ossification, secondary centers of ossification, epiphyseal plate, articular cartilage, marrow cavity.

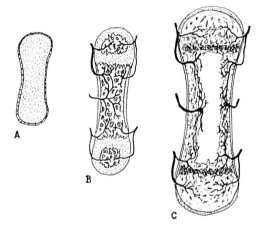

2. True or false? Upon completing the process of endochondral ossification, at about the age of 25, long bones cease to grow or change within the body___

3. Describe a set of circumstances that might result (a) in the decalcification of certain bones, and (b) in the thickening of certain bones.

   a. _____

   b. _____

VII. Bone Modeling and Remodeling (170)

Complete the following table.

What	Where	When	How and Why
Modeling			Growth of new layers of bone, resorption of old layers, in response to hormonal control
	Throughout adult life	At specific locations on the bone called foci	

VIII.   The Physiological Functions of Bones in Maintaining Homeostasis (171)

1. List the three minerals for which bones serve as the body's main storehouse.

    a. _____

    b. _____

    c. _____

2. What kind of bone cells control the storage and release of calcium? _____

3. Describe the composition and function of hemopoietic tissue. _____

4. Where in the skeletal system does one find red bone marrow? _____

5. Describe the roles of PTH and calcitonin in maintaining calcium homeostasis in blood.

    _____

6. Name two minerals and three vitamins necessary for proper bone growth.

    a. _____

    b. _____

    c. _____

    d. _____

    e. _____

IX. The Effects of Aging on Bones (173)

1. What three changes in bone tissue result in an increased brittleness in the bones of older people?

    a. _____

    b. _____

    c. _____

X. Developmental Anatomy of the Skeleton (173)

1. What are the two main divisions of the skull?

    a. _____

    b. _____

2. What kind of bone (based on embryonic derivation) are the frontal and parietal bones and portions of the temporal and occipital bones? _____

3. Which portion of the neurocranium develops from a cartilaginous model? _____

4. What bones form the viscerocranium? _____

5. Which bone of the appendicular skeleton does not develop from a cartilaginous model?

    _____

6. How many primary ossification centers does a vertebra have, and to what part(s) of the vertebra do they pertain? _____

XI. When Things Go Wrong (174)

Complete the following table.

Disease	Cause(s)	Symptom(s)
Osteogenesis imperfecta		Localized pain and tumors are signs of malignancy in this rare form of cancer
	Staphylococcus aureus and other bacteria	
Paget's disease		
	Vitamin D deficiency	
		Bones become so porous that they crumble under ordinary stress of moving about

Key Terms

accessory bone 159
bone-lining cell 164
bone marrow 159
canaliculi 161
central (Haversian) canal 161
compact bone tissue 161
diaphysis 159
endochondral ossification 166
epiphyseal (growth) plate 159
epiphysis 159
flat bone 157
intramembranous ossification 164

irregular bone 157
lacuna 161
lamellae 161
long bone 157
metaphysis 159
modeling 170
osseous tissue 157
ossification 164
osteoblast 163
osteoclast 163
osteocyte 163
osteogenic cells 163

osteon 161
osteoporosis 175
perforating canal 161
periosteum 159
remodeling 170
sesamoid bone 157
short bone 157
spongy bone tissue 161
trabeculae 161

Post Test

Multiple Choice

____ 1. What do these words have in common: vertebrae, trabeculae, lacunae,
scapulae, Caesar?

   a. they are all anatomical terms
   b. they all start with a capital letter
   c. they all end with "ae"
   d. the diphthong, ae, has the sound "-ee" in all of them.
   e. none of the above

____ 2. What anatomical entity is as strong
as iron, light as wood, a storage
depot for calcium and phosphorus, a
major site for hematopoiesis, support for the body, and found only in
vertebrate animals?

   a. blood
   b. integument
   c. bone
   d. cartilage
   e. all of the above

____ 3. Bones are usually classified by
shape, such as

   a. flat
   b short
   c. all of the above
   d. long
   e. irregular

____ 4. Supporting elements that require
only limited movement, such as in
the wrist and/or ankle, yet are almost
completely covered with articular
surfaces, describes

   a. long bones
   b. irregular bones
   c. short bones
   d. sesamoid bones
   e. flat bones

____ 5. A small bone that is embedded in
the tendon of one of the large
muscles of the thigh

   a. femur
   b. hip bone
   c. patella
   d. humerus
   e. accessory bone

____ 6. The smallest bones in the human
body

   a. middle ear bones
   b. distal finger bones
   c. sutural bones
   d. sesamoid bones
   e. pneumatic bones

____ 7. The Wormian bones are

   a. long bones
   b. pisiform bones
   c. accessory bones
   d. none of the above
   e. composite bones

____ 8. The membrane that lines most of the
internal cavities of bones is known
as the

   a. synovial membrane
   b. endosteum
   c. epiphyseal membrane
   d. perichondrium
   e. periosteal membrane

____ 9. Covering the outer surface of bones is the

a. endosteum
b. peripheral membrane
c. periosteum
d. both a and c
e. circumosteum

____ 10. The membrane covering the outer surface of bones is often attached to the bone by collagenous fibers called

a. articular fibers
b. Wormian fibers
c. endosteal fibers
d. Sharpey's fibers
e. matrical fibers

Matching

____ 11. Openings in compact bone that allow the entry and exit of blood vessels

____ 12. The part of a mature long bone that is not covered by periosteum

____ 13. As important as calcium and phosphorus to the physical and mechanical properties of compact and spongy bone

____ 14. The spongy material found occupying most of the space between the tables of certain cranial bones

____ 15. Cylinders of compact bone

a. osteons
b. nutrient foramina
c. collaginous fiber network
d. diploë
e. articular surface of epiphyses

____ 16. Multinuclear giant cells
____ 17. Cells that dissolve bone
____ 18. Cells that synthesize and secrete osteoid
____ 19. Small, spindle-shaped cells with a high mitotic potential; found in the deepest layers of periosteum
____ 20. Cells that act as pumps and move calcium and phosphate into and out of bone tissue
____ 21. The main cells of fully developed bones
____ 22. The progenitors of osteoclasts
____ 23. Derived from osteoblasts, they keep the matrix in a stable and healthy state by maintaining its mineral content
____ 24. White blood cells
____ 25. Can be transformed into osteoblasts during healing

a. osteogenic (osteoprogenitor) cells
b. osteoblasts
c. osteocytes
d. osteoclasts
e. monocytes

___ 26. The concentric layers of bony tissue that make up the osteons of compact bone

___ 27. Right angle branches of the central canals of compact bone

___ 28. Longitudinally disposed cylinders of compact bone

___ 29. Nonossified spaces that contain the osteocytes

___ 30. Structures in cancellous bone that form along the lines of greatest pressure or stress

a. Volkmann's canals
b. Haversian system
c. trabeculae
d. lamellae
e. lacunae

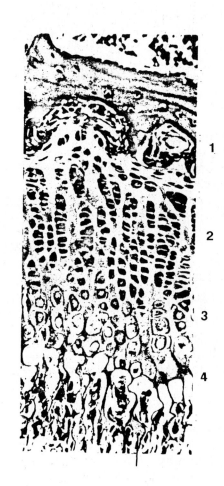

Identify each of the zones in the figure above.

31. Zone #1 _____

32. Zone #2 _____

33. Zone #3 _____

34. Zone #4 _____

35. What is shown next to and below Zone #4? _____

Multiple Choice

____ 36. Intramembranous ossification is the process by which bone develops
a. from connective tissue membranes
b. directly within a connective tissue network
c. directly within a cartilaginous model
d. indirectly from a focus of osteogenic cells
e. from calcium carbonate and sodium phosphate

____ 37. Bones that develop by intramembranous ossification are
a. the long bones of the legs
b. the long bones of the arms
c. flat bones like the shoulder blade (scapula)
d. flat bones of the face and skull
e. the hip bones

____ 38. The first rapid phase of intramembranous bone development begins around the second month of prenatal life with the appearance of
a. a layer of mesenchyme forming around the brain
b. a plentiful supply of blood vessels in the mesenchyme between the brain and scalp
c. a ring-shaped ossification center
d. lacunae enclosing osteocytes and osteo-blasts

____ 39. In the beginning of intramembranous ossification, the mesenchymal cells differentiate from osteogenic cells into
a. osteoid-secreting osteocytes
b. osteoid-secreting osteoblasts
c. osteoid-forming osteoclasts
d. osteoids
e. both b and c above

____ 40. As the osteoblasts cause calcium salt deposits to form a spongy bone matrix, they become entrapped in spaces called
a. lacunae
b. canaliculi
c. Haversian canals
d. Volkmann's canals

____ 41. As osteocytes, these entrapped cells make contact with other osteocytes and osteoblasts through
a. nutrient foramina
b. bony trabeculae
c. mineral deposits
d. gap junctions

____ 42. Collaginous fibrils that become deposited on the trabeculae entrap blood vessels and combine to form
a. blood-forming bone marrow
b. secondary osteogenic rings
c. centers for endochondral ossification
d. compact bone

____ 43. The osteoblasts on the surface of the spongy bone tissue form the
a. endosteum
b. periosteum
c. perichondrium
d. endochondrium
e. both b and c

____ 44. The replacement of the cartilaginous skeleton of the fetus by bone tissue constitutes what is known as
a. intramembranous ossification.
b. extramembranous ossification
c. endogenous bone replacement
d. endochondral ossification
e. periosteal ossification

___ 45. Chondrocytes in the middle of a
cartilage model trigger a chemical
reaction in the matrix by secreting

a. calcium salts
b. calcium and phosphate
c. calcium phosphatase
d. alkaline phosphatase
e. cartilage spicules

___ 46. The cartilage becomes mineralized
to form spicules, upon which the
osteoblasts lay down

a. and die
b. an osteoid matrix
c. spongy bone
d. cancellous bone
e. ossification centers

___ 47. The diffusion of nutrients to the
chondrocytes becomes blocked
during the early stages of endochondral ossification by mineralization of
the osteoid matrix, and the
chondrocytes

a. die
b. become transformed to pluripotent cells
c. form columns, the trabeculae
d. divide in their lacunae

___ 48. Pluripotent cells, which line the
cavities in the degenerating cartilage
model, differentiate into

a. osteoblasts
b. osteocytes
c. osteoblasts and osteocytes
d. osteoclasts
e. all of the above

___ 49. As the interior of the cartilage model
breaks down and is replaced by
spongy bone, the outside covering
membrane becomes transformed
into the

a. primary center of ossification
b. secondary center of ossification
c. diaphysis
d. periosteum

___ 50. While the main shaft of the skeletal
element is rapidly being transformed
to bone, the two ends become
invaded by osteogenic cells. These
elements constitute the

a. metaphyses
b. epiphyses
c. epiphyseal plates
d. diaphyses
e. all of the above

___ 51. Cells that are crucial to the remodeling of bone are the

a. osteoclasts
b. ostocytes
c. osteoblasts
d. all of the above

Matching

___ 52. A bone tumor
___ 53. A faulty calcification process
___ 54. A bone marrow infection
___ 55. "Adult rickets"

a. rickets
b. osteomalacia
c. osteomyelitis
d. osteosarcoma

___ 56. Stimulates endochondral ossification.

___ 57. Depresses nervous system, leading to muscular weakness

___ 58. Directly affects body's acid-base balance

___ 59. Increases neuronal membrane permeability, causing muscle spasms

___ 60. Manifested by loss of appetite and constipation

___ 61. Speeds up bone remodeling

a. hypopcalcemia
b. hypercalcemia
c. phosphate metabolism
d. PTH
e. thyroxine

Integrative Thinking

1. The x-ray pictures of two patients at a local clinic were not properly marked, so the orderly was not sure which was which. They were both x-rays of the wrist: one of an elderly woman, and the other of a 9-year old child. One of the pictures showed that the bones were very thin, and the other showed a dark line across the radius and ulna near the ends of those bones. Which picture belonged to whom, and how can you tell?

2. What skeletal abnormalities would you predict for a woman having depressed thyroid function and elevated parathyroid hormone (PTH) secretion?

3. A young man is being treated with both radiation and chemotherapy for cancer. The physician advised him that there might be serious side-effects, because these treatments could destroy the stem cells in his myeloid tissues. What effect might this have on the young man's ability to fight infectious diseases? Why?

Your Turn

Here are some correct answers to multiple-choice questions. Make up the questions plus the incorrect answers.

1. "Haversian systems but not trabeculae"

2. "Red marrow, but not yellow marrow"

3. "Chondrocytes, but not osteoblasts"

4. "The production of osteoblasts from osteoprogenitor cells in the periosteum"

7 The Axial Skeleton

Active Reading

I. General Features and Surface Markings of Bones (180)

1. How many bones does the typical adult have? _____

2. What parts of the skeleton make up the axial skeleton? _____

3. What parts make up the appendicular skeleton? _____

4. Name three functions of the skeleton.

    a. _____

    b. _____

    c. _____

5. Define the following terms. Indicate whether each is a process, a depression, or an opening.

    a. condyle _____

    b. crest _____

    c. fissure _____

    d. fossa _____

    f. trochlea _____

    g. trochanter _____

    h. tubercle _____

    i. tuberosity _____

    j. sulcus _____

    k. sinus _____

    l. facet _____

    m. spine _____

    n. foramen _____

    o. notch _____

    p. head _____

    q. canal _____

6. What types of forces give rise to processes and other features on the bone's surface? ____

_____

II. Divisions of the Skeleton (180)

1. What are the major components of the axial skeleton?_____

2. Name the major components of the appendicular skeleton. _____

3. Technically speaking, is the hyoid bone a part of the skull? _____

III. The Skull (184)

1. How many bones make up the skull? _____

2. The skull can be divided into what two main regions?

    a. _____

    b. _____

3. What are the functions of the two regions?

    a. _____

    b. _____

4. Name the associated bones of the skull? _____

5. Name the bones of the cranial skull, marking paired bones with the letter P. _____

_____

6. What are the main foramina of the skull? _____

7. Describe the location of the following sutures.

    a. coronal _____

    b. lambdoidal _____

    c. sagittal _____

    d. squamous _____

8. Identify the fontanels at the junctions of the following bones.

    a. frontals, parietals _____

    b. parietals, occipital _____

    c. parietal, occipital, temporal _____

    d. parietal, frontal, sphenoid, temporal _____

9. Which fontanels persist in an adult's skull? _____

10. Identify the eight bones of the cranial skull._____

_____

11. What is the calvaria, and what bones make up the calvaria? Also, what bones make up the base?_____

_____

12. Complete the table below.

Feature	Location and/or Function
	Shell of the forehead
Orbital plate	
	Eyebrow ridge
Supraorbital foramen	

13. Emissary veins connecting the superior sagittal venous dural sinus inside the skull with scalp veins outside the skull pass through foramina in the_____ bones.

14. Describe the locations and functions of the following features of the occipital bone.

    a.   foramen magnum _____

    b.   occipital condyles_____

    c.   hypoglossal canals _____

    d.   atlantooccipital joint_____

    e.   external occipital crest _____

    f.   external occipital protuberance _____

15. What are the four parts of each temporal bone?

    a. _____

    b. _____

    c. _____

    d. _____

16. On the figure below, locate the following bones: occipital, parietal, temporal, frontal, sphenoid, zygomatic, maxilla, mandible.

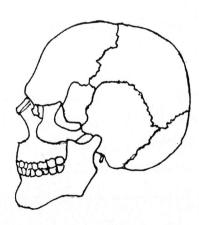

17. On the figures below, locate the following parts of the ethmoid bone: cribriform plate, orbital plate, perpendicular plate, crista galli, middle nasal concha, ethmoidal air cells.

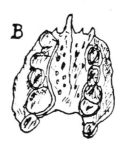

18. List the paranasal sinuses and name their main function(s). _____

_____

19. In what bones do the paranasal sinuses lie? _____

20. In the sutures of the calvaria lie the _____ bones.

21. How many bones are there in the facial skull? _____

22. On the figure below, locate the following bones and openings: mandible, maxilla, zygomatic bone, nasal bone, lacrimal bone, vomer, zygomaticofacial foramen, infraorbital foramen, nasal cavity, mental foramen.

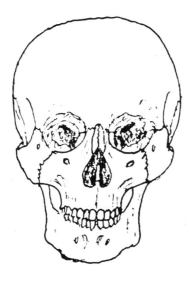

Match the following pairs.

____ 23. Symphysis menti
____ 24. Palatine bones
____ 25. Lacrimal sac
____ 26. Inferior nasal conchae

a. thin bony plates that increase the surface area of the nasal mucosa
b. fusion point for the left and right halves of the mandible
c. depression in the lacrimal bone which collects excess tears from the surface of the eye
d. posterior part of the hard palate, parts of floors and walls of nasal cavity, and floor of orbit

27. On the figure below, locate the following: body of mandible, ramus, angle, mandibular notch, coronoid process, condylar process, mandibular foramen, alveolar process, and mental foramen.

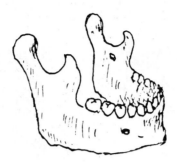

28. What two functions does the hyoid bone serve?

   a. _____

   b. _____

29. What are the three ossicles?

   a. _____

   b. _____

   c. _____

30. Where are they located? _____

IV. The Vertebral Column (201)

1. How many vertebrae make up the vertebral column? _____

2. Where are the intervertebral disks located? _____

3. List the five regions of the vertebral column, from superior to inferior. Give the number of separate bones in each region.

   a. _____

   b. _____

   c. _____

   d. _____

   e. _____

4. What are the location and shape of the two primary curves?

   a. _____

   b. _____

5. What are the location and shape of the two secondary curves?

   a. _____

   b. _____

6. Complete the following table.

Vertebral Element	Location	Function
Spinous process	Posterior wall of vertebra	Attachment site for vertebral muscles
Transverse processes		
Superior articular processes		
Inferior articular processes		
Vertebral foramen		
Intervertebral foramen		
Vertebral body		

7. What are the vertebrae of the neck called? _____

8. What is their function? _____

9. In what structural ways do they differ from the other vertebrae? _____

10. How does the first cervical vertebra, called the _____, differ from the other cervicals? _____. Where is its body? _____. With what does it articulate superiorly _____ and posteriorly? _____

11. Name and describe the second cervical vertebra. _____

12. What characterizes the injury known as "whiplash"? _____

13. Name and describe the seventh cervical vertebra. _____

14. What are the anatomical features of the thoracic vertebrae? _____

15. What are the costal facets, and what is their function? _____

16. Label the anatomical parts of the vertebra illustrated below.

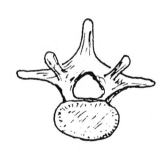

17. Which vertebrae are the largest and strongest? _____

18. What special anatomical features characterize the lumbar vertebrae? _____

_____

19. How many vertebral bodies make up the sacrum? _____

20. What is the function of the sacrum? _____

21. What structures make up the sacroiliac joints? _____

22. Define or describe each of the following.

    a. sacral canal _____

    b. sacral hiatus _____

    c. sacral promontory _____

    d. lumbosacral angle _____

23. Describe the coccyx. _____

V. The Thorax (209)

1. What skeletal elements comprise the thorax? _____

2. What are the functions of the thorax? _____

3. Label all parts of the figure below.

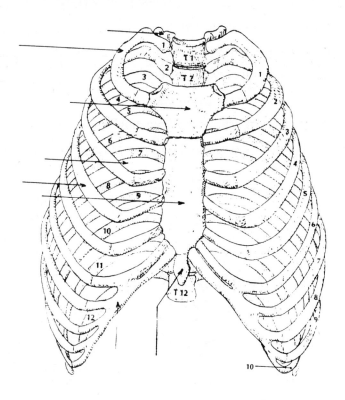

VI. When Things Go Wrong (212)

1. Describe (a) the physical nature, (b) the cause(s), and (c) the treatment of herniated disks.

    a. _____

    b. _____

    c. _____

2. Describe the condition cleft palate. _____

3. Describe the anomalies (a) microcephalus and (b) hydrocephalus.

    a. _____

    b. _____

4. Define or briefly describe (a) spina bifida, (b) kyphosis, (c) lordosis, and (d) scoliosis.

    a. _____

    b. _____

    c. _____

    d. _____

Key Terms

appendicular skeleton 184
atlas 204
axial skeleton 184
axis 204
cervical vertebrae 204
coccyx 208
cranial skull 184
ear ossicles 201
ethmoid bone 197
facial skull 184
fontanel 184
foramen 184
foramen magnum 190
frontal bone 189
hyoid bone 200
inferior nasal conchae 199

intervertebral disks 201
lacrimal bones 199
lumbar vertebrae 206
mandible 200
maxillary bones 199
nasal bones 199
occipital bone 190
palatine bones 199
paranasal sinuses 197
parietal bones 189
process 180
ribs 212
sacrum 208
skull 184
sphenoid bone 195

spinal curves 202
sternum 209
sutural bones 197
suture 184
temporal bones 190
thoracic vertebrae 204
thorax 209
vertebra 201
vertebral arch 204
vertebral body 204
vertebral canal 204
vertebral column 201
vertebral foramen 204
vomer 199
zygomatic bones 199

Post
Test

Multiple Choice

___ 1. The two major divisions of the human skeleton are the
  a. skull and vertebral column
  b. spinal column and appendages
  c. axial and appendicular skeletons
  d. skull and axial skeleton
  e. all of the above

___ 2. The skull can be divided into the
  a. cranium and mandible
  b. cranial and facial components
  c. calvaria and base components
  d. occipital and frontal components
  e. upper and lower halves

___ 3. The skull is lightened by cavities called
  a. fontanels
  b. facial fissures
  c. nasal turbinals
  d. nasal fossa
  e. paranasal sinuses

___ 4. Except for the mandible, ear ossicles, and the hyoid bone, the bones of the skull are joined by
  a. fontanels
  b. Wormian bones
  c. spinous processes
  d. sutures
  e. ridges

___ 5. All but one of the following are names of fontanels. The one exception is the
  a. parietal
  b. mastoid
  c. occipital
  d. sphenoidal
  e. frontal

Matching

___ 6. A wide, prominent ridge, often on the long border of a bone
___ 7. A hammer-shaped, rounded process
___ 8. A relatively narrow, tubular channel, opening to a passageway
___ 9. Natural opening through or into a bone
___ 10. A canal
___ 11. A groove or cleft
___ 12. A small, roughly rounded process
___ 13. A rounded, knuckle-shaped projection
___ 14. A small, flat surface
___ 15. A "lip ridge"

a. fissure
b. meatus
c. malleolus
d. tubercle
e. facet
f. crest
g. foramen
h. condyle

___ 16. A deep indentation, especially on a
bone's border
___ 17. A shallow, depressed area
___ 18. Medium-sized, roughly rounded,
elevated process
___ 19. A process that supports
___ 20. A curved, hornlike protuberance
___ 21. The projecting part of a bone,
generally smooth
___ 22. A deep depression on the border of a
bone
___ 23. A sharp, elongated process
___ 24. A shallow, depressed area
___ 25. A grooved surface serving as a sort
of pulley

a. sustentaculum
b. spine
c. fossa
d. cornu
e. trochlea
f. notch
g. tuberosity
h. eminence

___ 26. A cranial suture located between
right and left parietal bones
___ 27. A suture between frontal and parietal bones of the skull
___ 28. The anterior fontanel, also known as
the ___ fontanel
___ 29. Not a name of a cranial suture
___ 30. A suture located between parietal
and occipital bones

a. coronal
b. frontal
c. squamous
d. parasagittal
e. sagittal
f. lambdoidal

___ 31. Located between the occipital and
the two parietals
___ 32. Posterolateral location
___ 33. At junction of frontal, parietal,
temporal, and sphenoid bones
___ 34. At junction of parietal, occipital, and
temporal bones
___ 35. Located between the angles of the
parietals and the frontal

a. anterior fontanel
b. posterior fontanel
c. anterolateral fontanel
d. mastoid fontanel

Match the following, based on the figure below.

_____ 36.
_____ 37.
_____ 38.
_____ 39.
_____ 40.
_____ 41.
_____ 42.
_____ 43.

a. frontal bone
b. maxilla
c. parietal bone
d. sphenoid
e. temporal bone
f. occipital bone
g. zygomatic bone
h. mandible

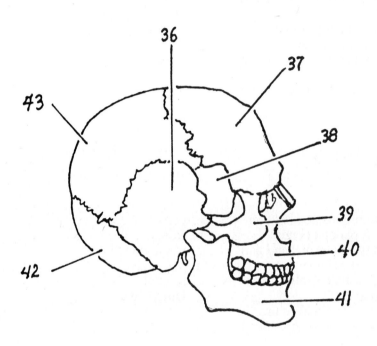

Matching

_____ 44. Cribriform plate
_____ 45. Supraorbital ridge
_____ 46. Foramen magnum
_____ 47. Crista galli
_____ 48. Petrous portion
_____ 49. Hypoglossal canals
_____ 50. Major part of the middle cranial fossa
_____ 51. Contains the auditory ossicles
_____ 52. Forms the superior sides, the roof, and part of the back of the skull
_____ 53. Supports the mastoid process
_____ 54. Sella turcica

a. frontal bone
b. parietal bone
c. occipital bone
d. temporal bone
e. sphenoid bone
f. ethmoid bone

Matching (Warning: some of the items may not be used)

_____ 55. The smallest facial bones, rectangu-
        lar shape
_____ 56. Has the mental foramen
_____ 57. Form the posterior part of the roof of
        the mouth
_____ 58. Infraorbital foramen
_____ 59. Thin, scroll-shaped bones
_____ 60. The "plowshare" bone that forms
        part of the nasal septum

a. inferior nasal conchae
b. vomer
c. palatine bones
d. maxillae
e. zygomatic bones
f. lacrimal bones
g. mandible
h. hyoid

Identify the various features illustrated in the following figure.

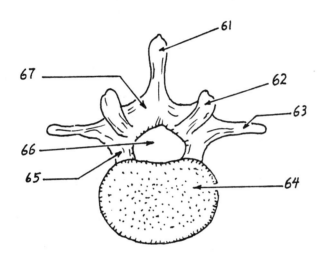

61. _____

62. _____

63. _____

64. _____

65. _____

66. _____

67. _____

Identify the structures illustrated in the following figure.

___ 68.	a. body
___ 69.	b. superior articular facet
___ 70.	c. superior demifacet
___ 71.	d. inferior articular process
___ 72.	e. intervertebral notch
___ 73.	f. lamina
___ 74.	g. transverse costal facet
___ 75.	h. spinous process
___ 76.	i. pedicle
___ 77.	

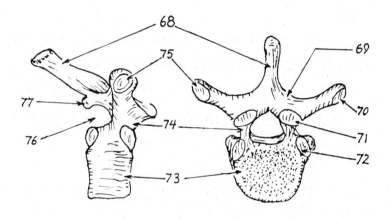

Short Answer

___ 78. What is the normal number of cervical vertebrae?
___ 79. What is the normal number of thoracic vertebrae?
___ 80. What is the normal number of lumbar vertebrae?
___ 81. What is the normal number of vertebrae that are fused to form the sacrum?

Matching

___ 82. Has auricular surfaces	a. atlas
___ 83. The seventh cervical	b. intervertebral foramina
___ 84. Formed by the juxtaposition of superior and inferior notches	c. forked spinous processes
	d. intervertebral disks
___ 85. Articulates with occipital condyles	e. axis
___ 86. Occur only in cervical vertebrae	f. sacrum
___ 87. Possesses the dens	g. vertebra prominens
___ 88. Composed principally of fibrocartilage	h. transverse foramina
___ 89. Characteristic of the 3rd to 6th cervicals	
___ 90. Characterized by four transverse lines on the anterior surface	

Matching

____ 91. Connects ribs R8 to R10 to sternum
____ 92. Attached to sternum indirectly
____ 93. Possesses a xiphoid process
____ 94. Has no connection to the breastbone
____ 95. Connection between rib and vertebral column

a. sternum
b. true rib
c. false rib
d. floating rib
e. costal cartilage
f. costal facet

Integrative Thinking

1. What major bone in the facial skull articulates directly with every other bone, except one, in the facial skull?_____. What is the exception?

2. Joe has a pituitary tumor. The surgeon will enter through the roof of the mouth. What bone(s) must she go through in order to reach the tumor?

3. Carmella was in an automobile accident and suffered whiplash. What are the various symptoms she might expect as a consequence, ranging from the mildest to the most severe? What anatomical structures would be directly affected?

4. Ms. Chu's physician, Dr. Steffanopolis, needs a sample of her cerebrospinal fluid for diagnostic purposes. How will he obtain it?

Your Turn

1. Write up a case history on a severe incidence of herniated disk, including symptoms, a detailed description of the injury, and the involvement of other tissues or organs. Also describe the surgical procedures to correct or ameliorate the problem.

2. Compose a set of multiple-choice questions (three or four) that deal with some aspect of this chapter.

8 The Appendicular Skeleton

Active Reading

I. The Upper Extremities (Limbs) (219)

1. In addition to the arms, forearms, and hands, what are the components of the upper extremities? _____

2. In addition to the legs and feet, what are the components of the lower extremities? _____

   _____

3. By what joint do the upper extremities attach to the axial skeleton? _____

4. What are the bones of the pectoral girdle? _____

5. Why is the shoulder a frequent site for dislocation injuries? _____

6. Complete the following table.

End of Clavicle	Articulates With	Joint Formed	Ligament Involved
Sternal		Sternoclavicular	
	Acromion process		Coracoclavicular

7. What is the point of attachment on the clavicle for the costoclavicular ligament? _____

   _____

8. What is the point of attachment on the clavicle for the coracoclavicular ligament? _____

   _____

9. On the figures below, locate the following: body of the scapula, spine, supraspinous fossa, infraspinous fossa, acromion, glenoid fossa, coracoid process, medial border, superior border, superior angle, inferior angle, lateral border.

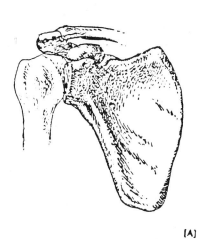 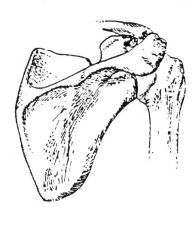

[A]   [B]

Match the following pairs.

___ 10. Forearm
___ 11. Wrist
___ 12. Hand
___ 13. Fingers

a. radius and ulna
b. metacarpal bones
c. carpal bones
d. phalanges
e. humerus

Describe the position and function of the following features of bones of the arm.

14. greater and lesser tubercles _____

15. anatomical neck _____

16. surgical neck _____

17. deltoid tuberosity _____

18. radial groove _____

19. trochlea _____

20. capitulum _____

21. lateral epicondyle _____

22. medial epicondyle _____

23. radial and coronoid fossa _____

24. olecranon fossa _____

25. What components of the ulna form the semilunar, or trochlear, notch? _____

26. Where on the ulna does the head of the radius articulate? _____

27. What are the locations and functions of the styloid processes? _____

_____

28. What are the location and function of the ulnar notch? _____
_____

29. When the arm is in anatomical position, which of the two bones of forearm occupies the medial position? _____

30. On the figure below, locate the following: scaphoid, lunate, triquetrum, pisiform, trapezium, trapezoid, capitate, hamate, metacarpals I to V, proximal phalanx, middle phalanx, distal phalanx, IP joint, DIP joint, PIP joint.

31. Which carpal bone is most prone to fracture? _____

II. The Lower Extremities (226)

1. How many bones make up the upper (____) and lower (____) extremities?

Match the following pairs.

____ 2. Thigh   a. tarsal bones
____ 3. Leg   b. metatarsal bones
____ 4. Ankle   c. tibia and fibula
____ 5. Foot   d. femur
____ 6. Toes   e. phalanges

7. In two languages, what bones make up the pelvic girdle? _____ (English), _____ (Latin)

8. What are the three functions of the pelvis?

a. _____

b. _____

c. _____

9. Contrast the range of motion and degree of support offered by the pelvic and pectoral girdles. _____

10. Through what part of the pelvis does a baby pass during childbirth? _____

11. How does the male pelvis differ from the female pelvis?

    a. _____

    b. _____

    c. _____

12. On the figure below, locate the following: ilium, pubis, ischium, acetabulum, sacroiliac joint, symphysis pubis, sacrococcygeal joint, greater (false) pelvis, lesser (true) pelvis, sacral promontory, iliac crest, anterior superior iliac spine, anterior inferior iliac spine, superior ramus of ischium, inferior ramus of ischium, superior ramus of pubis, inferior ramus of pubis, oburatur foramen.

13. The greater sciatic notch is a feature of the (a) ilium, (b) ischium, or (c) pubis?___ On which part of the os coxa is the lesser sciatic notch?

14. Describe the positions and functions of the following.

    a. head of femur _____

    b. neck of femur _____

    c. greater trochanter _____

    d. lesser trochanter _____

    e. fovea capitis _____

    f. shaft of femur _____

    g. linea aspera _____

    h. medial condyle _____

    i. lateral condyle _____

    j. medial epicondyle _____

    k. lateral epicondyle _____

    l. patellar surface _____

15. How does the knee "lock"? _____

16. What kind of a bone is the patella? _____

17. What is the function of the patella? _____

18. True or false? The apex is the top of the patella and the base is the bottom. ___

19. Which bone of the lower leg bears the weight of the body? _____. Which bone aids in the motion of the ankle? _____. Which bone occupies the lateral position? _____

20. Describe the function of the tibial tuberosity. _____

21. What parts of the tibia articulate with the femur? _____

22. With what bone do the medial malleolus and the lateral malleolus articulate? _____

23. Describe the position and function of the following.

    a. fibular notch_____

    b. tibiofibular joints _____

    c. interosseous membrane _____

24. What bones does the talocrural joint involve? _____

25. True or false? The ankle is the most stable when it is dorsiflexed.___

Match the following pairs:

___ 26. Talus	a.	most lateral tarsal bone
___ 27. Calcaneus	b.	border metatarsals I–III
___ 28. Cuboid bone	c.	ankle bone
___ 29. Navicular bone	d.	heel bone
___ 30. Cuneiform bones	e.	lies at head of talus

31. On the figure below, locate the following: talus, calcaneus, navicular, cuboid, medial cuneiform, intermediate cuneiform, lateral cuneiform, metatarsals I–V, proximal phalanx, middle phalanx, distal phalanx.

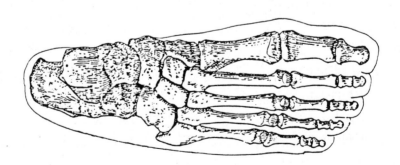

32. What function do the arches of the foot serve? _____

33. What bones make up the medial longitudinal arch? _____

34. What bones make up the lateral longitudinal arch? _____

35. How are the arches held in place? _____

36. What is the function of the plantar calcaneonavicular ligament? _____

III. When Things Go Wrong (236)

1. Contrast the effects on the spinal cord of compression and extension fractures of the vertebral column. _____

2. Describe the process by which a fracture heals. _____
_____

3. What is a bunion? _____

4. What are shin splints, and how do they develop? _____

5. In what sports are athletes prone to develop shin splints? _____

Key Terms

appendicular skeleton 219
arch 235
carpal bones 224
clavicle 219
femur 230
fibula 232
humerus 222
ilium 226
ischium 226

lower extremity 226
metacarpal bones 224
metatarsal bones 235
oss coxae 226
patella 232
pectoral girdle 219
pelvic girdle 226
pelvis 226, 230
phalanges 227, 235

pubis 230
radius 223
scapula 219
shoulder girdle 219
tarsal bones 232
tibia 232
ulna 223
upper extremity 219

Post Test

Matching

___ 1. Calcaneus
___ 2. Carpals
___ 3. Humerus
___ 4. Metatarsus
___ 5. Talus
___ 6. Clavicle
___ 7. Pubis
___ 8. Patella
___ 9. Phalanges
___ 10. Femur
___ 11. Radius
___ 12. Ulna
___ 13. Acetabulum
___ 14. Coracoid process
___ 15. Fibula
___ 16. Metacarpus
___ 17. Scapula
___ 18. Ischium
___ 19. Tibia
___ 20. Ilium

a. pectoral girdle
b. pelvic girdle
c. arm
d. forearm
e. wrist
f. hand
g. fingers
h. thigh
i. leg
j. ankle
k. heel
l. foot
m. kneecap

Matching

___ 21. Glenoid fossa
___ 22. Pisiform bone
___ 23. Acromion
___ 24. Conoid tubercle
___ 25. Ulnar notch
___ 26. Radial notch
___ 27. Deltoid tuberosity
___ 28. Olecranon fossa
___ 29. Scaphoid
___ 30. Semilunar or trochlear notch

a. clavicle
b. scapula
c. humerus
d. ulna
e. radius
f. carpus

Label the figure below, using the numbers for reference.

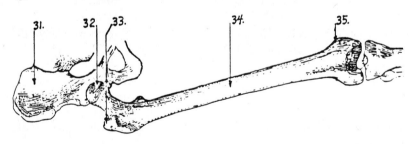

31._____ 34._____

32._____ 35._____

33._____

Similarly label the following figure.

36. _____   39. _____

37. _____   40. _____

38. _____

Integrative Thinking

1. The most commonly broken bones are the clavicle (leading the list), the neck of the femur, the head of the humerus, the distal end of the radius and/or ulna, and the wrist. Why do you think that the clavicle is especially vulnerable? What is peculiar about the posture of a person with a broken clavicle as he or she walks?

2. It is common for elderly people, especially women, to break the neck of the femur. These people are very surprised when this happens because, as they will tell you, "When I was a kid I thought nothing of jumping out of a tree", or "I could do the Lindy Hop all night long, and then jump down the stairs to the car for a ride home." Now they say, "I was just walking along when for no reason at all I fell down and broke my hip." What they find hard to believe is that maybe their "hip" (neck of the femur) broke first, and then they fell. How do you account for this? What role might osteoporosis have played?

3. A common reflex that comes into play when one pitches forward in a fall is to throw out one or both hands to break the fall. This usually results in a sprain of the wrist, but sometimes in a fracture of one of the wrist bones. Pretend you are falling in this way, and throw out, say, your left hand. Stop! Look at your hand! How is it oriented? Now look at the figure of the bones of the hand and wrist in your textbook. Which wrist bone do you think will be affected most by the fall?

Your
Turn

1. The dictionary defines mnemonics as a system or device for helping the memory. It is used extensively by medical students who are trying to memorize an anatomical series, such as the cranial nerves ("On Old Olympus's Towering Top A Fat Armed German Vaults And Hops", for Olfactory, Optic, Oculomotor, etc.). For your turn, make up a mnemonics for the bones of the wrist.

2. Make up a mnemonics for the bones of the tarsus.

9 Articulations

Active Reading

I. Classification of Joints (243)

1. What is another name for a joint? _____

2. Are all joints movable? _____

3. What are the two methods by which joints are classified? _____

Match the following triplets.

___, ___ 4. Synarthrosis   a. limited motion   i. functional classification
___, ___ 5. Cartilaginous   b. generally limited motion   ii. structural classification
___, ___ 6. Diarthrosis   c. freely movable
___, ___ 7. Fibrous   d. generally freely movable
___, ___ 8. Synovial   e. no motion
                f. synovial

II. Fibrous Joints (243)

1. How are bones connected in a fibrous joint? _____

2. What element common to other kinds of joints does a fibrous joint lack? _____

3. Complete the following table.

Joint	Structure	Degree of Movement	Example
Suture			
Syndesmoses			
Gomphoses			

III. Cartilaginous Joints (243)

1. In what two ways are cartilaginous joints united?

    a. _____

    b. _____

2. Do cartilaginous joints have a joint cavity? _____

3. How much movement do cartilaginous joints permit? _____

4. Complete the following table.

Joint	Structure	Example 1	Example 2
		Symphysis pubis	Intervertebal disks
Synchondrosis			

5. What is the primary function of most synchondroses? _____

IV. Synovial Joints (246)

1. Define the four essential elements of a synovial joint.

    a. synovial cavity _____

    b. articular cartilage _____

    c. articular capsule _____

    d. ligaments _____

2. Label the structures shown in the following diagram of the knee joint.

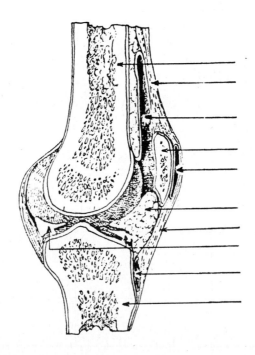

___ 3. Synovial membranes

    a. contain pads of fat
    b. help fill up space in the synovial cavity
    c. reduce friction within the joint
    d. secrete synovial fluid
    e. all of the above

___ 4. Articular disks

    a. secrete added synovial fluid
    b. hold the synovial joint together
    c. act as shock absorbers in some joints
    d. reduce the likelihood of joints fitting together too evenly
    e. all of the above

___ 5. The meniscus is an example of

    a. articular disk
    b. fibrocartilage
    c. complete partition
    d. all of the above

___ 6. When overstressed, a ligament will

    a. contract
    b. stretch
    c. tear
    d. spasm
    e. all of the above

7. Describe the structure and function of the bursae. _____

8. In the wrist, palm, and fingers, the bursae have been modified to form _____ .

9. In the diagram below, identify the following: bone, ligament (fibrous capsule), articular cartilage, periosteum, articular disk, synovial cavity, synovial membrane.

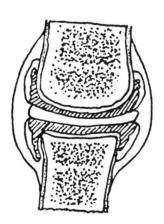

V. Movement at Synovial Joints (248)

1. What three factors limit movement at synovial joints?

    a. _____

    b. _____

    c. _____

Match the following functions with their actions and their opposing actions.

___ 2. Raising a body part	a. abduction	i. extension
___ 3. Movement of limb away from midline	b. plantar flexion	ii. depression
	c. flexion	iii. adduction
___ 4. Radius rotates diagonal to ulna	d. elevation	iv. dorsiflexion
___ 5. Movement of sole of foot inward	e. protraction	v. opposition
___ 6. Extension of foot at ankle	f. pronation	vi. eversion
___ 7. Thumb returns to anatomical position	g. inversion	vii. supination
	h. reposition	viii. retraction
___ 8. Forward movement of jaw		
___ 9. Angle between bones decreases		

10. Define the following terms.

    a. palmar flexion _____

    b. hyperextension _____

    c. circumduction _____

    d. rotation _____

11. In the following table, check off which type of movement(s) correspond(s) to which type of joint.

Joint	Flexion	Abduction	Rotation
Uniaxial			
Tiaxial			
Multiaxial/triaxial			

VI. Types of Synovial Joints (248)

Complete the following table. In each description, include whether the range of motion is uniaxial, biaxial, or multiaxial.

Joint	Description	Example
Hinge		
Pivot		
Condyloid		
Gliding		
Saddle		
Ball and socket		

VII. Nerve Supply and Nutrition of Synovial Joints (252)

True or false.

1. Any given joint may be supplied by the branches of several nerves ____

2. Many sensory nerve fibers terminate as nerve endings in the fibrous capsules, ligaments, and synovial membranes of the joints ____

3. Most of the innervation of joints is by motor (efferent) nerve fibers ____

4. What is the source of nourishment for the articulating cartilages of synovial joints? ____

   _____

5. How is synovial fluid circulated in the joint capsule? _____

6. What is the nature of the lymphatic drainage of synovial joints, if there is any? _____

VIII. Description of Some Major Joints (253)

Describe the following joints. In each description, include the points of articulation, the classification(s) of the joint (e.g. ball and socket), and the type(s) of movement permitted (e.g., protraction).

1. Temporomandibular _____

   _____

2. Glenohumeral _____

   _____

3. Coxal _____

   _____

4. Tibiofemoral _____

   _____

5. Label the following structures shown in the diagram below, which is a knee joint from which most structures have been removed: femur, tibia, anterior cruciate ligament, posterior cruciate ligament, and the tibial epiphysis.

IX. Developmental Anatomy of Joints (263)

1. What is mesenchyme? _____

Match the following pairs.

____ 2. Mesenchyme gives rise to capsule and ligaments peripherally, disappears centrally, and forms the synovial membrane in between.

____ 3. Mesenchyme differentiates into dense, fibrous connective tissue.

____ 4. Mesenchyme differentiates into hyaline cartilage or fibrocartilage.

a. fibrous
b. cartilaginous
c. synovial

X. The Effects of Aging on Joints (263)

1. What two effects does aging have on joints?

    a. _____

    b. _____

2. What are two diseases that often affect joints with age?

    a. _____

    b. _____

3. What two effects does aging have on ligaments?

    a. _____

    b. _____

4. Why is a ruptured vertebral disk less likely to occur in old age than in earlier years? ____

    _____

XI. When Things Go Wrong (263)

1. Name the following two conditions:

    a. inflammation of the joints _____

    b. inflammation of the bursae _____

Match the following pairs:

____ 2. Degeneration of articular cartilage; affects weight-bearing joints

____ 3. Inflamation of synovial membrane; autoimmune disease leading to bony ankylosis

____ 4. Combination of uric acid with sodium to form crystal deposits in articular cartilage

a. gouty arthritis
b. osteoarthritis
c. rheumatoid arthritis

5. Complete the following table.

Condition	Cause(s)	Symptom(s)	Treatment
Ankylosing spondylitis			
	Tearing of ligaments following sudden wrenching of joint		
		Pain of the TMJ muscles	

6. Describe the benefits of CPM therapy. _____

_____

Key Terms

abduction 251
adduction 251
amphiarthrosis 243
arthroscopy 261
articular capsule 246
articular cartilage 246
articulation 243
axis of rotation 251
ball-and-socket joint 252
biaxial joint 248
bursa 247
cartilaginous joint 243
circumduction 251
condyloid joint 248
coxal (hip) joint 256
depression 251
diarthrosis 243
dorsiflexion 250
elevation 251

eversion 250
extension 250
fibrocartilaginous disk 246
fibrous joint 243
flexion 250
glenohumeral joint 256
gliding joint 252
gomphosis 243
hinge joint 248
hyaline cartilage 246
hyperextension 250
inversion 250
ligament 247
meniscus 246
multiaxial joint 248
opposition 250
palmar flexion 250
pivot joint 248
plantar flexion 250

pronation 250
protraction 251
reposition 250
retraction 251
rotation 251
saddle joint 252
supination 250
suture 243
symphysis 246
synarthrosis 243
synchondrosis 243
syndesmosis 243
synovial cavity 246
synovial fluid 246
synovial joint 246, 248
temporomandibular joint 253
tendon sheath 247
tibiofemoral joint 258
uniaxial joint 248

Post Test

Multiple Choice

____ 1. Joints are classified according to their

    a. degree of movement
    b. physical location
    c. mineral composition
    d. fibrocartilage composition
    e. capsular shape

____ 2. According to structural classification, the kinds of joints are

    a. diarthroses
    b. fibrous
    c. cartilaginous
    d. synovial
    e. b, c, and d

____ 3. Immovable joints are known as

    a. amphiarthroses
    b. diarthroses
    c. synarthroses
    d. syndesmoses
    e. all of theabove

____ 4. A fibrous joint that appears as a peg
and socket, like the relation between
a tooth and the jawbone, is a

a. suture
b. gomphosis
c. synostosis
d. syndesmosis
e. synarthrosis

____ 5. A "skull type" joint is more technically known as a

a. synarthrosis
b. diarthrosis
c. amphiarthrosis
d. suture
e. threaded joint

Matching

____ 6. The elbow joint
____ 7. The carpometacarpal joint of the
thumb
____ 8. The joint between the atlas and axis
____ 9. Most knuckle joints
____ 10. Hip joint
____ 11. Acromioclavicular joint
____ 12. Shoulder joint
____ 13. Ankle
____ 14. Joint between articular processes of
vertebrae
____ 15. Proximal radioulnar joint

a. hinge
b. pivot
c. condyloid
d. gliding
e. saddle
f. ball-and-socket

____ 16. Forms a more or less complete
transverse partition
____ 17. Folds of tissue that line the joint
cavity of a diarthosis
____ 18. The four essential structures of this
kind of joint are the joint cavity, the
articular cartilage, the articular
capsule, and ligaments
____ 19. A primary cartilaginous joint
____ 20. May act as a shock absorber between bones at a synovial joint
____ 21. The midline joint between the pubic
portions of the paired hip bones

a. synchondrosis
b. symphysis
c. synovial
d. articular disk
e. meniscus
f. synovial membrane

____ 22. Fibrous thickenings of the articular
capsule joining two bones
____ 23. Collagenous fibers that join muscle
to bone
____ 24. Flattened sacs filled with synovial
fluid
____ 25. Tend to tear, rather than stretch, if
subjected to excessive stress
____ 26. Long, cylindrical sacs filled with
synovial fluid

a. ligaments
b. bursae
c. tendon sheaths
d. tendons

Multiple Choice

___ 27. Lysosome-containing cells that are common in synovial membranes
    a. plasma cells
    b. erythrocytes
    c. phagocytes
    d. osteocytes
    e. osteoblasts

___ 28. Consist of plasma dialysates (filtrates), globular proteins, and mucin (hyaluronic acid)
    a. ligaments
    b. synovial membranes
    c. synovial fluids
    d. interosseous tendons
    e. fibrocartilage sheaths

___ 29. The "torn cartilages" that plague athletes in rough, contact sports
    a. articular ligaments
    b. menisci
    c. synovial membranes
    d. bursae
    e. tendon (synovial) sheaths

___ 30. Inward movement of the sole of the foot
    a. dorsiflexion
    b. hyperextension
    c. pronation
    d. eversion
    e. inversion

Matching

___ 31. Keeping time with your foot while standing
___ 32. Raising your hand as a signal to stop
___ 33. Turning your hand inward, palm facing downward
___ 34. Making a fist
___ 35. Increasing the angle at a joint
___ 36. Straightening the elbow
___ 37. Slapping your thigh while standing
___ 38. Twirling a lariat at your side
___ 39. Turning your hand so that the radius and ulna are parallel to one another
___ 40. A parking attendant signaling cars to turn right to enter the lot

    a. flexion
    b. extension
    c. dorsiflexion
    d. circumduction
    e. supination
    f. pronation
    g. abduction
    h. adduction

___ 41. A symphysis between the manu-
brium and the body of the sternum

___ 42. A gliding and rotational joint between the ribs and the transverse
processes of thoracic vertebrae

___ 43. Effects a gliding movement between
successive costal cartilages of ribs
5–9

___ 44. Joint between the body of the
sternum and the xiphoid process

___ 45. A syndesmosis between the cartilage
of rib 1 and the sternum

a. costovertebral joint
b. sternocostal joint
c. interchondral joint
d. manubriosternal joint
e. xiphisternal joint

Integrative Thinking

1. In baseball every team member, except the pitcher, can expect to play in every game, barring illness or injury. Pitchers work in rotation, with two or three days of rest between assignments. What reasons can you give to account for this kind of concern?

2. As people age, they appear to become shorter in stature and they don't stand up as straight as they used to. How do you account for this?

3. Why are ligaments so important in certain joints, such as the shoulder, elbow, hip, and knee, when they appear to resist the very movements that the joints are designed to make?

4. What is the TMJ syndrome and what may bring it about?

Your Turn

By now your small study groups are well organized and thoroughly imbued with a spirit of cooperation and intellectual curiosity. As a group, go over the Suggested Readings in your textbook. Let each member of the group select one of the references for Chapter 9 to report on at the next meeting of your group. Compose a note to your instructor, telling him or her what the group thought of the reference — whether or not it was interesting, if it was informative, if you think it should be required reading, or whatever you feel is appropriate. Who knows? You might get brownie points for your efforts!

10 Muscle Tissue

Active Reading

Introduction

1. What are the three types of muscle tissue?

   a. _____

   b. _____

   c. _____

2. What are the basic physiological properties of muscle tissue?

   a. _____

   b. _____

   c. _____

3. What are the three general functions of muscle tissue?

   a. _____

   b. _____

   c. _____

I. Anatomy of Skeletal Muscle (272)

1. Why is skeletal muscle called "striated muscle"? _____

2 Why is skeletal muscle said to be "voluntary"? _____

3 What is meant by "muscle tone"? _____

4. What are fascia? _____. What are the two major forms, and where are they located?

   a. _____

   b. _____

Match the following pairs.

___ 5. Encloses a muscle   a. perimysium
___ 6. Encloses a fascicle   b. epimysium
___ 7. Encloses a muscle fiber c. endomysium

8. Of what type of tissue are tendons and aponeuroses composed? _____. What is their function? _____

___ 9. The blood vessels that supply blood   a. arteries
       to the individual muscle fibers are   b. veins
                                              c. capillaries
                                              d. all of the above

10. Label the structures shown in the diagram below.

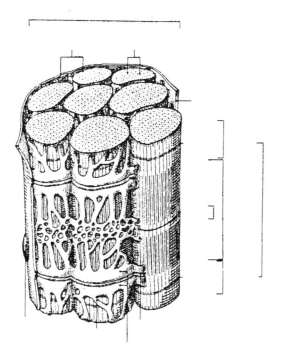

11. What is a motor unit? _____

12. Define the following terms.

   a. neuromuscular junction _____

   b. motor end plate _____

   c. synaptic gutter _____

   d. subneural cleft _____

   e. synaptic cleft _____

13. Label the diagram below.

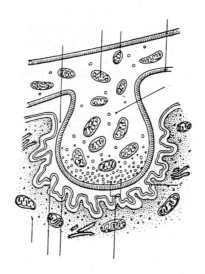

Refer to the figure below, and label the following statements as true (T) or false (F).

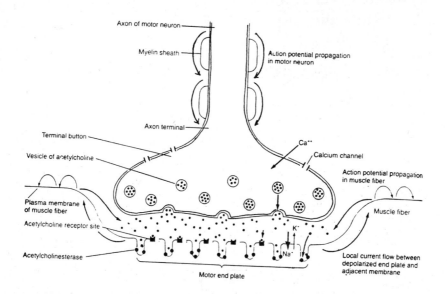

____ 14. A nerve impulse, propagated along the axon of a motor neuron, triggers the opening of voltage-gated channels on the axon terminal.

____ 15. Once opened, these ion channels allow sodium ions to enter the axon terminal and permit potassium ions to leave.

____ 16. The increased Ca^{2+} concentration in the the axon terminal triggers the exocytosis of acetylcholine from the synaptic vesicles in the axon terminal and into the synaptic cleft.

____ 17. Acetylcholine diffuses across the synaptic cleft, attaching to specific receptor sites on the motor end plate of the sarcolemma.

____ 18. The binding of acetylcholine to these receptors triggers the opening of sodium and potassium channels. Because the flow of potassium out of the muscle fiber is greater than the inflow of sodium, a further polarization at the motor end plate results. This polarization reduces any possible end-plate potential.

____ 19. The end-plate potential depolarizes adjacent sarcolemma. If the sarcolemma reaches its threshold potential, it initiates an action potential, which propagates over the surface of the entire muscle fiber and continues into the fiber via the transverse tubules. The action potential triggers the release of Ca2+ ions from the sarcoplasmic reticulum.

____ 20. Acetylcholinesterase then breaks down the acetylcholine into acetate and choline, allowing the ion channels in the motor end plate to close and the motor end plate to return to its resting potential.

____ 21. If the nervous stimulation is insufficient to reach the muscle fiber's threshold potential, the muscle fiber will nevertheless contract, albeit with less intensity.

22. Contrast the absolute refractory period with the relative refractory period. _____

_____

23. What happens to the muscle fiber's threshold potential during the relative refractory period? _____

24. What three proteins make up an actin myofilament?

   a. _____

   b. _____

   c. _____

25. Label the structures shown in the diagram below.

   a. _____

   b. _____

   c. _____

   d. _____

   e. _____

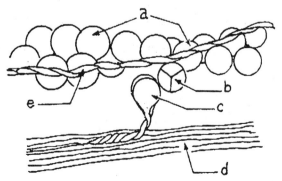

II. Physiology of Muscle Contraction (278)

1. What effect does calcium released from the sarcoplasmic reticulum have on the troponin complexes? _____

2. According to the sliding-filament model, why does the formation of myosin cross bridges lead to the pulling of the actin myofilaments toward the direction of the H zone?

   _____

3. Why does the pulling of the actin myofilaments toward the direction of the H zone result in muscle contraction? _____

4. During muscle relaxation, what effect does the transport of calcium ions back to the sarcoplasmic reticulum have on myosin cross bridges? _____

5. Describe the role of creatine phosphate during prolonged muscle contraction. _____

   _____

6. What happens in each of the four sequences of muscle relaxation?

   a. _____

   b. _____

   c. _____

   d. _____

7. What are the sources of ATP needed for muscle contraction? _____

III. Types of Muscle Contraction (281)

1. Contrast isotonic and isometric muscle contraction. _____

2. Name and describe the various types of muscle contraction.

   a. _____

   b. _____

   c. _____

   d. _____

3. What is the all-or-none principle? _____

IV. Grading (Varying Strength) of Muscle Contraction (284)

1. What are the main factors affecting the grading of muscle contraction? _____

   _____

2. What factors determine or characterize the ideal resting length of muscle fibers? _____

   _____

3. What provides the body with its major source of heat for temperature regulation? _____

   _____

4. Describe how the following phenomena are effected.

    a. anaerobic respiration _____

    b. muscle fatigue _____

    c. oxygen debt _____

5. What effects are produced by the different types of exercise? _____

_____

V. Fast-Twitch and Slow-Twitch Muscle Fibers (286)

1. What is the essential difference between slow-twitch (type I) and and fast-twitch (type II) muscle fibers? _____

2. What are the differences between slow and fast fibers and what are some examples of each? _____

VI. Smooth Muscle (287)

1. In what four systems is smooth (involuntary) muscle most frequently found?

    a. _____

    b. _____

    c. _____

    d. _____

2. What are two main characteristics of smooth muscle?

    a. _____

    b. _____

Associate the following features with either skeletal (SK), smooth (SM), or both (B) kinds of muscle.

3. The cells are called fibers. ___

4. Contains multiple nuclei. ___

5. Contains troponin. ___

6. Sarcoplasmic reticulum is poorly developed. ___

7. Myofibrils are very thin and arranged more randomly. ___

8. Cross bridges are formed between myofilaments. ___

VII. Contraction of Smooth Muscle (289) .

1. How are single-unit smooth muscle fibers arranged? _____

2. What two types of contractions take place in single-unit smooth muscle fibers?

    a. _____

    b. _____

3. How does multiunit smooth muscle differ from single-unit smooth muscle? _____

_____

4. What are the three sources of calcium in smooth muscle?

  a. _____

  b. _____

  c. _____

5. Describe the multistep process by which the input of calcium causes smooth muscle tissue to contract. _____

6. What are five inputs that trigger calcium movement within smooth muscle cells?

  a. _____

  b. _____

  c. _____

  d. _____

  e. _____

VII. Cardiac Muscle (290)

1. Where in the body is cardiac muscle found? _____

2. Contrast the presence of mitochondria in skeletal and cardiac muscle tissue. _____

  _____

3. What are the three functions of intercalated disks?

  a. _____

  b. _____

  c. _____

4. How many nuclei does a typical cardiac muscle cell have? _____

5. Describe what happens to cardiac muscle when it does not receive nervous stimuli. ___

  _____

6. Contrast the sarcoplasmic reticulum and calcium sensitivity of skeletal and cardiac muscle tissues. _____

7. Why is the heart not susceptible to tetanic contraction? _____

IX. Developmental Anatomy of Muscles (292)

___ 1. From what type of cells are muscles   a. endodermal
       derived?   b. mesodermal
                                                    c. ectodermal
                                                    d. mesenchyme
                                                    e. both b and d

2. What are the three layers of a somite?

  a. _____

  b. _____

  c. _____

3. From which of these three layers does most skeletal muscle develop? _____

X. The Effects of Aging on Muscles (293)

1. What are six effects of age on muscles?

    a. _____

    b. _____

    c. _____

    d. _____

    e. _____

    f. _____

2. Why is it important that elderly people continue to be as mobile as possible? _____

   _____

XI. When Things Go Wrong (293)

1. Describe the healing processes in the three types of muscles following injury.

    a. skeletal_____

    b. smooth _____

    c. cardiac _____

2. Complete the following table.

Condition	Cause(s)	Symptom(s)
Disuse atrophy		
Denervation atrophy		
Myopathy		
Dystrophy		

3. What holds promise as a treatment for DMD?_____

4. What is myasthenia gravis?_____

5. Why is the bacterium *Clostridium tetani* more likely to multiply in deep puncture wounds than in superficial wounds? _____

   _____

Key Terms

acetylcholine 277
actin 273
all-or-none principle 282
cardiac muscle tissue 271, 290
cross bridge 279
endomysium 272
epimysium 272
excitation-contraction coupling 279
fascia 272
fascicle 272
fast-twitch muscle 286
intercalated disk 290
isometric contraction 284
isotonic contraction 283
motor end plate 275
multiunit smooth muscle 288

muscle fatigue 286
muscle fiber 273
myofibril 273
myofilament 273
myosin 273
myotome 292
neuromuscular junction 275
oxygen debt 286
pacemaker cell 288
perimysium 272
refractory period 281
sarcolemma 273
sarcomere 273
sarcoplasm 273
sarcoplasmic reticulum 273
single-unit smooth muscle 288

skeletal muscle tissue 271, 272
sliding-filament model 278
slow-twitch muscle 286
smooth muscle tissue 271, 287
somite 292
striated muscle tissue 271, 272
synaptic cleft 277
tetanus 282
transverse tubule 273
treppe 283
tropomyosin 279
troponin 279
twitch 281
voluntary muscle 272

Post Test

Multiple Choice

____ 1. The basic physiological property of muscle tissue is

a. excitability
b. extensibility
c. refractility
d. contractility

____ 2. The major general function of muscle tissue is

a. metabolism
b. movement
c. respiration
d. secretion

____ 3. Another major general function of muscle tissue is

a. heat production
b. energy production
c. respiration
d. reproduction

____ 4. The individual cells of skeletal muscle are known as

a. muscle fibers
b. myofibrils
c. contractile units
d. myofiolaments

___ 5. Skeletal muscle fibers are

   a. multicellular
   b. multistranded
   c. multinucleate
   d. branched

Matching

___ 6. Heavily striated
___ 7. Finely striated
___ 8. Nonstriated
___ 9. Has motor end plates
___ 10. Characterized as voluntary
___ 11. Cells mononucleate and spindle-shaped
___ 12. Associated with internal organs, such as blood vessels

   a. skeletal muscle
   b. smooth muscle
   c. cardiac muscle

Multiple Choice

___ 13. Which is the correct descending order of size?

   a. myofilament>myofibril>myosin>actin
   b. myofibril>myofilament>actin>myosin
   c. myosin>actin>myofibril>myofilament
   d. myofilament>myofibril>actin>myosin
   e. myofibril>myofilament>myosin>actin

___ 14. What is the fundamental unit of muscle contraction?

   a. sarcolemma
   b. sarcomere
   c. sarcoplasm
   d. sarcoplasmic reticulum

___ 15. The I bands of skeletal muscle striations represent

   a. actin alone
   b. myosin alone
   c. actin and myosin together
   d. absence of both actin and myosin

Label the diagram below.

___ 16. endomysium
___ 17. fascicle
___ 18. perimysium
___ 19. epimysium
___ 20. blood vessel

Label the diagram below.

___ 21. A band
___ 22. M line
___ 23. Z line
___ 24. I band
___ 25. H zone
___ 26. sarcomere

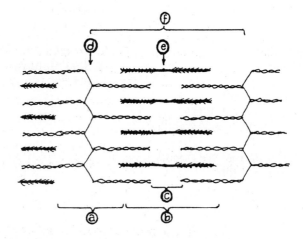

Matching

___ 27. The area on a skeletal muscle fiber along which a motor neuron ending makes contact

___ 28. The specialized structure that the muscle fiber membrane forms at its points of contact with the motor neuron endings

___ 29. The sarcolemma at these points of contact maximizes its area of contact by means of folds called

___ 30. A tiny gap separating the nerve ending and the sarcolemma at their point of contact

___ 31. Minute sacs in the cytoplasm of the nerve ending that contain the neurotransmitter substance

___ 32. The neurotransmitter substance
___ 33. The enzyme that deactivates the neurotransmitter
___ 34. Contains acetylcholine
___ 35. Site of action of acetylcholinesterase

a. motor end plate
b. motor unit
c. neuromuscular junction
d. acetylcholine
e. synaptic cleft
f. synaptic vesicle
g. subneural cleft
h. acetylcholinesterase

Multiple Choice

___ 36. The transmission of impulses from motor neuron endings is facilitated by openings in the neuron cell membrane of

a. sodium-potassium channels
b. voltage-gated channels
c. channels through which calcium ions flow out
d. channels through which calcium ions flow in

___ 37. Exocytosis of acetylcholine from vesicles in the terminal button is stimulated by

a. acetylcholinesterase
b. an increase in Ca^{2+} in the terminal button
c. a decrease in Ca^{2+} in the terminal button
d. the closing of the voltage-gated channels
e. the opening of the voltage-gated channels

___ 38. The neurotransmitter is secreted into the synaptic cleft, where it binds to

a. acetylcholinesterase
b. calcium ions
c. the channel proteins
d. specific acetylcholine receptor sites
e. actin and myosin

___ 39. The binding of acetylcholine causes

a. immediate generation of an action potential
b. the immediate release of Ca^{2+}
c. channels in the motor end-plate membrane to open
d. the muscle fiber to contract

___ 40. The channels that are opened in the preceding question facilitate the flow of

a. calcium ions back into the button
b. calcium ions back out of the button
c. Na^+ out of the muscle fiber
d. K^+ into the muscle fiber
e. Na^+ into and K^+ out of the muscle fiber

___ 41. The consequence of the flow of ions is

a. a drop in electrical charge across the muscle cell membrane
b. depolarization at the motor end plate
c. an increase in electrical charge across the membrane of the muscle fiber
d. creation of an end-plate potential
e. both b and d, which are the same

___ 42. The end-plate potential culminates in an action potential, which propagates over the surface of the entire muscle fiber and then enters the fiber via the transverse tubules. There the action potential causes the release of calcium ions from the

a. sarcoplasmic reticulum
b. motor end plate
c. muscle fiber
d. neurotransmitter vesicles in the button

___ 43. The result of the preceding is that the calcium ions stimulate all the myofibrils in the muscle cell

a. to secrete acetylcholine
b. to secrete acetylcholinesterase
c. to relax
d. to contract all at the same time

___ 44. Regardless of the magnitude of the nervous stimulus, the muscle cell receiving the stimulus will contract

a. maximally or not at all
b. weakly or strongly, depending on the strength of the stimulus
c. a number of times, depending on the strength of the stimulus
d. weakly or strongly, depending on the amount of calcium ions received

___ 45. The interval of time during which a muscle that has already contracted will not contract again is known as its

a. refectory period
b. refractory period
c. recalcitrant period
d. refusal period

___ 46. The energy required for contraction comes directly from

a. creatinine sulfate
b. glucose
c. glycogen
d. ATP

Matching

___ 47. A muscle contraction without shortening
___ 48. A muscle contraction with shortening
___ 49. A momentary muscle contraction in response to a single stimulus
___ 50. A sustained contraction owing to a rapid series of stimuli
___ 51. A succession of contractions of increasing magnitude, usually characteristic of a rested muscle

a. twitch
b. treppe
c. tetanus
d. isometric
e. isotonic

Identify each of the three kinds of muscle cells and the definitive structures indicated in the following figure.

___ 52. skeletal muscle
___ 53. nucleus
___ 54. intercalated disk
___ 55. smooth muscle
___ 56. cardiac muscle

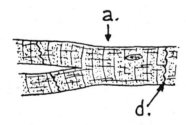 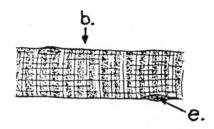 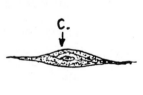

In the table below, rank the three types of muscle tissue as a (most pronounced, fastest, or longest), b (intermediate), or c (least pronounced, slowest, shortest, or lacking) in the following properties or characteristics.

	Property/Character	Skeletal	Cardiac	Smooth
57.	Shortening speed			
58.	Cross striations			
59.	Fatigue resistance			
60.	Sustaining a single contraction			
61.	Refractory period			
62.	Cell size			
63.	Amount of SR and T tubules			
64.	Mitochondria			

Matching

____ 65. Arrector pili muscles of the skin
____ 66. Exhibit tonic contractions
____ 67. Fibers individually innervated
____ 68. Single-unit muscles arranged in a ring structure
____ 69. Iris and ciliary muscles of the eye
____ 70. Capable of rhythmic contractions

a. single-unit type of smooth muscle
b. multiunit type of smooth muscle
c. both single-unit and multiunit
d. sphincters

Multiple Choice

____ 71. Cardiac muscle tissue appears

a. only in the heart
b. in the heart and lungs
c. in the heart and the aorta
d. in the heart and sinus venosus

____ 72. Muscle cells of the heart are joined end-to-end by

a. so-called "tight junctions"
b. fused cell membranes
c. so-called "intercalated disks"
d. fused Z lines

____ 73. Tetanization of cardiac muscle is avoided by

a. prolonged contraction
b. controlled calcium release
c. an extended refractory period
d. the SA-node, the "pacemaker" of the heart

Integrative Thinking

1. In most cases of poliomyelitis the motor neuron cell bodies in the spinal cord are killed, and they cannot regenerate. What effect does this have on the skeletal muscles that normally are innervated by these neurons, and what can be done about it?

2. The muscle fibers of the heart are branched. They are arranged in a loose, rather than tight, spiral and are joined end-to-end by special junctions that permit an action potential to pass from cell to cell. What is the advantage of these kinds of arrangements to the functioning of the heart?

Your Turn

1. Let the people in your study group take turns doing this: Stand in a doorway (one that is not too wide) and, with your arms at your side, palms inward, your eyes closed, push the backs of both hands against the door frame as hard as you can (without bursting a blood vessel). Keep pushing for about 10 seconds. Then, letting your arms relax and fall to your sides, step back out of the doorway, keeping your eyes closed. Just stand there, arms relaxed by your side, for 10 seconds or so. Now open your eyes. What happened? How do you explain it? Was the contraction of your arm muscles while you were doing this in the doorway isometric or isotonic?

2. Let each member of your study group make a schematic drawing of a skeletal muscle sarcomere as it would appear in, for some members, (a) a relaxed state; for others in (b) a partially contracted state; for still others, in (c) a fully contracted state; and finally (d) in a partially stretched state. Do not label them. Pass the completed sketches around, and after everybody has examined them all, spread them out on a table or desk and arrange the best of each state of contraction in a series: relaxed, partially contracted, fully contracted, and partially stretched. Which one would contract with the greatest force if stimulated in the condition shown?

11 The Muscular System

Active Reading

Introduction

1. By how much of their resting length can most muscles contract? _____

2. By how much of their resting length can most muscles stretch? _____

3. About how many skeletal muscles does a person have? _____

I. How Muscles Are·Named (300)

Match the following descriptions with their respective muscles.

___ 1. Obliquus
___ 2. Trapezius
___ 3. Levator scapulae
___ 4. Biceps
___ 5. Gluteus maximus
___ 6. Tibialis anterior
___ 7. Sternohyoid

a. shape, trapezoidal
b. size, largest
c. location, in front of tibia
d. attachment sites
e. heads of attachment, two
f. action, raises the scapula
g. direction of fibers, slanted

II. Attachment of Muscles (300)

1. What is the belly of a muscle? _____

2. Describe the structure and function of a tendon. _____

3. Compare an aponeurosis and a tendon. _____

4. Indicate whether the following features generally imply the origin (O) or insertion (I) of a muscle.

   a. proximal attachment _____

   b. bone that moves _____

   c. bone that is fixed _____

   d. distal attachment _____

III. Architecture of Muscles (301)

1. What determines a muscle's range of motion? _____

2. What determines a muscle's strength? _____

3. Into what bundles are muscle fibers grouped? _____

4. Complete the following table.

Muscle Type	Description	Example
Multipennate		
	Spindle shaped with thick belly	
		Flexor pollicis longus
Strap		
	Fascicles in circular shape around opening	
		Rectus femoris

IV. Individual and Group Actions of Muscles (000)

1. Complete the following table:

Type of Movement	Muscle Type
Flexion	
	Pronator
Rotation	
	Everter
Retraction	

2. Define the following types of muscles.

  a. agonist _____

  b. antagonist _____

  c. synergist _____

  d. fixator _____

V. Lever Systems and Muscle Actions (306)

1. What is a fulcrum? _____

2. What is a lever arm? _____

3. Provide two examples, one in the body and the other in the agricultural or construction industry (e.g, raising a wheelbarrow), of the following types of levers.

  a. first class _____

  b. second class _____

  c. third class _____

4. As leverage (mechanical advantage) increases, what happens to strength and to speed?

_____

VI. Specific Actions of Principal Muscles (307)

Muscles of Facial Expression

1. What source of innervation do all muscles of facial expression share? _____

2. Into what type of tissue do most muscles of the face insert?_____

3. Those lucky individuals who can wiggle their ears do so by using which muscles? _____

_____

4. Identify the muscles required to

    a. elevate eyebrows _____

    b. wink _____

    c. raise upper eyelid _____

    d. widen nasal aperture_____

    e. purse lips _____

    f. smile _____

    g. grin _____

5. Describe the location and function of the following muscles.

    a. platysma _____

    b. buccinator _____

    c. orbital _____

    d. occipitalis_____

Muscles That Move the Eyeball

6. What are the names and functions of the two oblique muscles of the eye?

    a. _____

    b. _____

Match the following pairs.

____ 7. Inferior rectus	a. rolls eye upward
____ 8. Medial rectus	b. rolls eye downward
____ 9. Lateral rectus	c. abducts eye
____ 10. Superior rectus	d. adducts eye

11. What does it mean that these muscles are extrinsic? _____

12. What is strabismus, and what muscles are at fault? _____

Muscles of Mastication

13. What source of innervation do the muscles of mastication share? _____

14. Complete the following table.

Muscle	Action	Origin	Insertion
		Coronoid process of mandible	
			Zygomatic bone
Lateral pterygoid			
Medial pterygoid			

15. What muscles interact to produce the side-to-side motion of chewing? _____

Muscles That Move the Hyoid Bone

16. Name and describe the actions of the four suprahyoid muscles.

a. _____

b. _____

c. _____

d. _____

17. Complete the following table.

Muscle	Origin	Insertion
Sternohyoid		
Stylohyoid		
Thyrohyoid		
Omohyoid — superior belly, inferior belly		

Muscles That Move the Tongue

18. Identify the muscles that produce the following actions.

a. alters shape of tongue _____

b. protrudes and depresses tongue _____

c. depresses tongue _____

d. retracts and elevates tongue _____

Muscles That Move the Head

___ 19. Flexion of the head is caused by contraction of

    a. bilateral semispinalis capitis
    b. unilateral splenius capitis
    c. bilateral sternocleidomastoid
    d. unilateral longissimus capitis

___ 20. Extension of the head is caused by contraction of

    a. bilateral sternocleidomastoid
    b. bilateral semispinalis, splenius, and longissimus capitis
    c. unilateral splenius capitis
    d. all of the above

___ 21. Which are the posterior neck muscles?

    a. splenius capitis
    b. semispinalis capitis
    c. longissimus capitis
    d. all of the above

Intrinsic Muscles That Move the Vertebral Column

22. Complete the following table.

Muscle	Origin	Insertion
Iliocostalis		
Longissimus		
Spinalis		

Match the following pairs.

___ 23. Iliocostalis cervicis
___ 24. Spinalis thoracis
___ 25. Longissimus lumborum
___ 26. Sacrospinalis

    a. erector spinae
    b. lateral cervical
    c. medial thorax
    d. middle lumbar

27. Which muscles flex the vertebral column? _____

28. Which group of muscles are superficial? _____

29. Which group of muscles are deep? _____

Muscles Used in Quiet Breathing

30. Describe the action of the following muscles.

    a. scalenes _____

    b. internal interchondrals _____

    c. internal intercostals _____

    d. external intercostals _____

    e. diaphragm _____

31. Describe the bucket-handle and pump-handle motions of the ribs. _____

_____

Muscles That Support the Abdominal Wall

32. Identify the following on the figure below: inguinal ligament, rectus sheath, linea alba, tendinous inscriptions, internal abdominal oblique, rectus abdominis, transversus abdominis, serratus anterior, external abdominal oblique, pectoralis major.

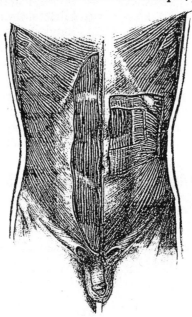

33. Where is the quadratus lumborum located? _____

34. What functions do the four muscles of the anterior and lateral walls serve? _____

_____

Muscles That Form the Pelvic Outlet

35. Identify the points of origin and insertion for the following muscles.

    a. pubococcygeus _____

    b. puborectalis _____

    c. coccygeus _____

    d. iliococcygeus _____

36. Which of the above muscles are not part of the levator ani muscle? _____

37. What are the five muscles of the perineum?

    a. _____

    b. _____

    c. _____

    d. _____

    e. _____

38. What muscle of the pelvic outlet is in a constant state of tonic contraction? _____

Muscles That Move the Shoulder Girdle

39. On the figure below, locate the following muscles: trapezius, levator scapulae, supraspinatus, major rhomboid, minor rhomboid, serratus posterior inferior, latissimus dorsi, longissimus thoracis, iliocostalis thoracis, deltoid, infraspinatus, teres minor, teres major.

40. When the glenoid fossa is directed upward, the scapula is said to rotate _____ .

Muscles That Move the Humerus at the Shoulder Joint

41. What are the four rotator cuff, or SITS, muscles?

    a. _____

    b. _____

    c. _____

    d. _____

42. What muscles are flexors?

    a. _____

    b. _____

    c. _____

    d. _____

_____ 43. Which muscle is not an extensor?   a. latissimus dorsi
    b. teres major
    c. deltoid
    d. supraspinatus

44. What type of motion do the pectoralis major and the coracobrachialis effect? _____

Muscles That Move the Forearm and Wrist

45. For each of the following muscles of the elbow joint, indicate whether the muscle is an extensor (E) or a flexor (F): biceps brachii (__), triceps brachii (__), brachialis (__), brachioradialis (__), anconeus (__).

46. Which two muscles pronate? _____, _____ Which two muscles supinate? _____, _____ Which type of motion is stronger? _____

Match the following muscles with the corresponding motion(s).

___ 47. Flexor carpi radialis
___ 48. Flexor carpi ulnaris
___ 49. Extensor carpi radialis longus
___ 50. Extensor carpi radialis brevis
___ 51. Extensor carpi ulnaris

a. extension
b. flexion
c. adduction
d. abduction
e. circumduction

Muscles That Move the Thumb

52. Describe the action of the following muscles:

    a. flexor pollicis longus _____

    b. opponens pollicis _____

    c. abductor pollicis longus _____

    d. adductor pollicis _____

    e. extensor pollicis longus _____

Muscles That Move the Fingers (Except the Thumb)

53. What muscles flex the metacarpophalangeal joints? _____

54. What muscles extend the metacarpophalangeal joints? _____

55. What action do the dorsal interossei effect? _____

56. What action do the palmar interossei effect? _____

57. When the hand is in anatomical position, what are the names of the five digits, progressing from medial to lateral?

    a. _____

    b. _____

    c. _____

    d. _____

    e. _____

58. Describe the actions of the following muscles.

    a. flexor digitorum superficialis _____

    b. flexor digitorum profundus _____

    c. abductor digiti minimi _____

59. Label all the muscles and tendons shown in the figures below.

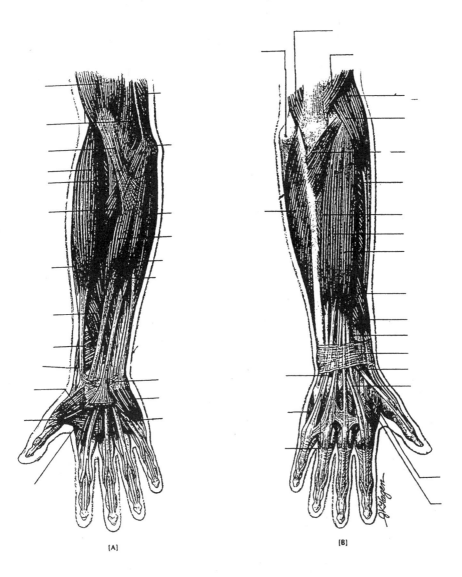

[A]   [B]

Muscles That Move the Femur at the Hip Joint

60. Complete the following table.

Muscle	Origin/Insertion	Action
Hamstring muscles		
Gluteus maximus		
Iliopsoas		
Tensor fasciae latae		
Adductor magnus		
Adductor longus		
Gluteus medius		

Muscles That Act at the Knee Joint

61. What are the four quadriceps femoris muscles?

    a. _____

    b. _____

    c. _____

    d. _____

62. What action do the quadriceps femoris muscles effect? _____

63. What are the three hamstring muscles?

    a. _____

    b. _____

    c. _____

64. Describe the action of the following three muscles.

    a. popliteus _____

    b. sartorius _____

    c. gracilis _____

64. Label all the muscles and tendons shown in the figures below:

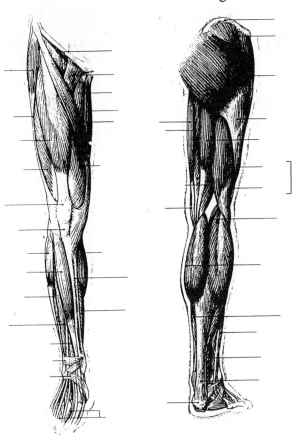

Muscles That Move the Foot

Match the following muscles with their respective actions.

___ 65. Soleus
___ 66. Gastrocnemius
___ 67. Tibialis anterior
___ 68. Tibialis posterior
___ 69. Peroneus longus
___ 70. Peroneus brevis
___ 71. Peroneus tertius

a. plantar flexion
b. dorsiflexion
c. inversion
d. eversion

Muscles of the Toes

72. Label the following figures as indicated by the label leaders.

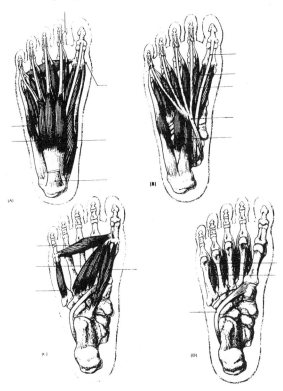

VII. When Things Go Wrong (345)

1. List the four injuries indicative of misuse of a computer keyboard, together with the accompanying causes.

Injury	Causes

2. Why and where are injections given intramuscularly? _____

3. What are the causes of hernias? _____

4. What is lateral epicondylitis? _____

5. What causes tension headaches? _____

Key Terms

agonist 303
antagonist 303
aponeurosis 300
belly 300
biceps brachii 325, 327
biceps femoris 335, 337
circular muscle 302
deltoid 325
erector spinae group 316
extrinsic muscle 310
fixator 303
force 306
fulcrum 306
fusiform muscle 302
gastrocnemius 337, 339
gluteus maximus 335

hamstring muscles 335
infraspinatus 325, 326
insertion 300
intercostal muscles 317
interossei 332, 340
latissimus dorsi 322, 325
leverage 306
lever arm 306
linea alba 318
origin 300
pectoralis major 325
pennate muscle 302
perineum 320
postural muscle 303
prime mover 303
quadriceps femoris 337

rectus abdominis 318
rotator cuff muscles 325, 326
sartorius 337
soleus 339
sphincter 308
sternocleidomastoid 315
strap muscle 302
subscapularis 325, 326
supraspinatus 325, 326
synergist 303
tendon 300
teres major 325
teres minor 325, 326
transversospinalis 316
trapezius 322
triceps brachii 325, 327

Post Test

Matching

___ 1. Movement toward midline of body
___ 2. Movement away from midline of body
___ 3. Turning around a longitudinal axis
___ 4. Extending foot at ankle toward sole
___ 5. Increasing angle at a joint
___ 6. Movement in a downward direction
___ 7. Turning foot so sole faces outward
___ 8. Decreasing angle at a joint
___ 9. Turning forearm so palm faces upward
___ 10. Reducing size of an orifice

a. rotation
b. supination
c. abduction
d. depression
e. adduction
f. extension
g. flexion
h. eversion
i. constriction
j. plantar flexion

___ 11. SITS
___ 12. Masseter
___ 13. Soleus
___ 14. Orbicularis oris
___ 15. Digastricus (ant. belly)
___ 16. Triceps brachii
___ 17. Gluteus medius
___ 18. Transversus abdominus

a. sphincter
b. plantar flexor
c. abductor
d. elevator
e. extensor
f. tensor (compressor)
g. depressor
h. rotator cuff

Match the following muscles with the class of lever that each utilizes — first, second, or third — as illustrated below (F=Force, W=Weight).

First Class Lever   Second Class Lever   Third Class Lever

___ 19. Gastrocnemius
___ 20. Quadriceps femoris
___ 21. Biceps brachii
___ 22. Splenius capitis
___ 23. Soleus
___ 24. Triceps brachii

a. first class
b. second class
c. third class

Matching

___ 25. A muscle that works together with a prime mover
___ 26. The prime mover
___ 27. The muscle that is primarily responsible for producing a particular movement
___ 28. A muscle that works counter to the prime mover
___ 29. A postural muscle

a. agonist
b. antagonist
c. synergist
d. fixator

___ 30. Longest muscle in the body
___ 31. Strongest for its size
___ 32. Weakest
___ 33. Largest, generally
___ 34. Smallest skeletal muscle
___ 35. Smooth muscle

a. stapedius
b. gluteus maximus
c. arrector pili
d. sartorius
e. masseter

Multiple Choice

___ 36. A favored site for an intramuscular injection is the
    a. temporalis
    b. vastus lateralis
    c. splenius
    d. gluteus medius
    e. psoas major

___ 37. The most common type of hernia is
    a. fetal
    b. inguinal
    c. gluteal
    d. ephemeral
    e. intestinal

___ 38. The operant fulcrum for what appears to be the only representative of a second-class lever in the human body
    a. ankle joint
    b. wrist joint
    c. patella of the knee
    d. shoulder joint
    e. ball of foot

___ 39. Most of the flexors of the digits are located in the
    a. arm
    b. forearm
    c. wrist
    d. hand
    e. fingers

___ 40. The muscle that grins in response to an amusing story
    a. risorius
    b. buccinator
    c. mentalis
    d. platysma
    e. palpebral

Integrative Thinking

1. The sartorius muscle is the longest muscle in the body, yet it takes its origin from the same place as the rectus femoris muscle, and the insertions of both are quite close to each other. How do you account for their difference in length?

2. How did the sartorius muscle get its name, which, by the way, is the Latin name for "tailor"?

3. In days long before jeeps or Sherman tanks, when mounted soldiers were much to be feared, enemy soldiers under the cover of darkness used to sneak into cavalry encampments and slash the hamstrings, or Achilles tendons, of the horses. Why did they do this cruel thing?

4. Why is the ankle joint not considered to be the fulcrum for plantar flexion of the foot?

Your Turn

Muscle Charades! Divide your study group or, if it's too small for this game, get together with another study group. (You will want to have two teams of at least four members each.) Have each team prepare as many cards as there are team members, and on them (the cards, not the players!) write the name of a particular activity and the names of the principal muscles that would be involved in executing that activity. Activities such as serving up a tennis ball, sipping a cup of tea, bowing to a Japanese diplomat, teeing off a golf ball, rinsing your hair in a wash basin, or even doing situps come to mind. Indicate whether the muscles are flexing, extending or fixating. When all the cards are complete, exchange them one at a time. One member of the opposing team would read to his/her teammates the muscles (and their actions) listed, and your team will time them to see how long it takes them to come up with the name or description of the particular activity. The shortest team time total takes the title!

12 Nervous Tissue: An Introduction to the Nervous System

Active Reading

Introduction

1. The process of _____ analyzes, combines, compares, and coordinates the messages of environmental change, received as _____ .

2. What are the three major regulatory systems of the body?

   a. _____

   b. _____

   c. _____

3. Of the three systems, which responds fastest? _____

4. By what means are stimuli processed by the nervous system? _____

I. Organization of the Nervous System (353)

1. What constitutes the central nervous system (CNS)? _____

2. What constitutes the peripheral nervous system (PNS)? _____

3. Into what two systems may the PNS be divided on a functional basis?

   a. _____

   b. _____

4. Name the two divisions into which each of these systems of the PNS is further divided.

   a. _____

   b. _____

II. Cells of the Nervous System (353)

1. What is a neuron? _____

2. What are the two important properties of neurons?

   a. _____

   b. _____

3. What are three principal parts of a neuron?

   a. _____

   b. _____

   c. _____

4. Do all neurons possess all three of these parts? _____

Match the following pairs.

____ 5. conduct nerve impulses toward the   a. Neurotubules
        cell body; extensions of the cell   b. Neurofilaments
        body; a neuron may have as many as   c. Dendrites
        200   d. Axons

____ 6. tube-like protein structures that
        transport substances between cell
        body and ends of processes

____ 7. conduct nerve impulses away from
        cell body; each cell generally has no
        more than one; range widely in
        length

____ 8. semirigid structures that provide
        skeletal framework for the axon of
        the neuron

9. Describe the following terms.

   a. collateral branch _____

   b. telodendria _____

   c. end bulb _____

   d. synapse _____

   d. myelin sheath _____

   e. Schwann cells _____

   f. oligodendrocytes _____

   g. neurilemma _____

10. How do Schwann cells and oligodendrocytes differ? _____

11. What are the gaps between myelin sheath units? _____

12. How do unmyelinated axons differ from myelinated axons in terms of conduction rate?

_____

13. Name and describe the four functional segments of neurons.

    a. _____

    b. _____

    c. _____

    d. _____

14. What is the difference between a tract and a nerve? _____

15. Distinguish between axoplasmic flow and axonal transport.

    a. _____

    b. _____

16. Describe the three functional classes of neurons.

    a. _____

    b. _____

    c. _____

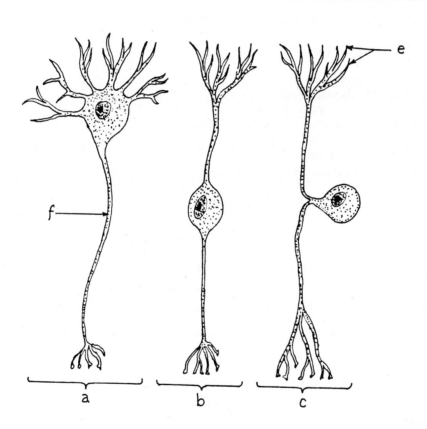

Identify the structures depicted in the diagrams above, according to the letters in the figure.

____ 17. unipolar neuron

____ 18. multipolar neuron

____ 19. bipolar neuron

____ 20. most sensory neurons of somatic PNS

____ 21. most neurons of brain and spinal cord

____ 22. axon

____ 23. dendrites

24. Name and explain the function of the four types of neuroglia.

    a. _____

    b. _____

    c. _____

    d. _____

25. True or false? Schwann cells may enable damaged peripheral nerves to regenerate. ___

26. What is the function of satellite cells? _____

27. What is N-CAM, and what role does it play in the development of the central nervous system? _____

III. Physiology of Neurons (263)

1. Why is a resting neuron said to be polarized? _____

2. What is the charge of the ICF relative to that of the ECF in a "resting" neuron, and what is that difference at any given point called? _____

3. What are the two major factors that contribute to the resting membrane potential?

    a. _____

    b. _____

4. How do the ECF and ICF compare in terms of ion concentration? _____

5. How does the flow of these ions across the membrane compare? _____

6. What is the approximate value, in millivolts, of the resting membrane potential? _____

7. What is the source of energy that "powers" the sodium-potassium pump? _____

8. What two ions have the most important effect on the resting membrane potential?

    a. _____

    b. _____

9. What is the ratio of sodium ions to potassium ions that the sodium-potassium pump transports? _____

10. What are the two types of channels through the neuron cell membrane?

    a. _____

    b. _____

11. By what means do open channels permit the flow-through of ions? _____

12. Describe the physiology of the three kinds of gated channels.

    a. _____

    b. _____

    c. _____

13. What is a stimulus that is strong enough to initiate a nerve impulse called? _____

14. What is the proper sequence of the following events? ___, ___, ___, ___, ___.

    a. A threshold stimulus causes gated channels to allow sodium to flow through the plasma membrane, giving the inner side of the membrane a positive charge relative to the outer side, and thus depolarizing the membrane.

    b. After each firing of the neuron, a refractory period of 0.5 to 1 msec takes place.

    c. The depolarization of one area of the plasma membrane causes voltage-gated sodium ion channels in adjacent areas to open. The successive triggering of voltage-gated channels causes an action potential, or nerve impulse, to move along the plasma membrane.

    d. The sodium-potassium pumps restore the original balance of ions across the cell plasma membrane, as well as the resting membrane potential. This repolarizes the plasma membrane.

    e. The outflow of K^+ ions while the voltage-gated channels are open may result in hyperpolarization.

15. Compare the all-or-none principle for neurons with what you have learned for muscle tissue. _____

16. According to the all-or-none principle, what might cause differences in perception between light stimuli and strong ones? _____

17. Nerve impulses pass along myelinated neurons by the process of _____ conduction.

18. Why do impulses travel so much faster along myelinated neurofibers than along unmyelinated fibers? _____

19. What factors determine the speed of impulse production? _____

20. What is a synapse? _____

Match the following.

  ___ 21. Neuron, muscle cell, or gland cell   a. presynaptic cell
  ___ 22. Leads toward the synapse   b. postsynaptic cell
  ___ 23. Leads away from the synapse
  ___ 24. Almost always a neuron

25. What are the three morphological classes of synapses?

    a. _____

    b. _____

    c. _____

26. What are the two functional classes of synapses?

    a. _____

    b. _____

27. What are connexons? _____

28. Explain the process by which a stimulus crosses an electrical synapse. _____

_____

_____

29. Explain the roles of the following in a chemical synapse.

    a. voltage-gated calcium ion channels _____

    b. calcium ions _____

    c. exocytosis _____

    d. neurotransmitter _____

    e. receptor protein sites _____

    f. chemically gated sodium ion and potassium ion channels _____

    g. acetylcholinesterase _____

30. What happens to neurotransmitters after a nerve impulse crosses a chemical synapse? ___

_____

31. What is a PSP? _____

Match the following pairs.

  ___ 32. Partial depolarizing
  ___ 33. Hyperpolarizing
  ___ 34. Effect on postsynaptic neuron of
          repeated firing of presynaptic neuron
  ___ 35. Summing of synaptic inputs from
          different presynaptic neurons
  ___ 36. Increases chances of generating an
          action potential in third neuron
          (postsynaptic excitation)
  ___ 37. Decreases chances that an action
          potential will have an excitatory
          effect at the synapse with a third
          neuron

a. spatial summation
b. temporal summation
c. excitatory postsynaptic potentials
d. inhibitory postsynaptic potentials
e. presynaptic inhibition
f. presynaptic excitation

38. What is the importance of inhibition? _____

39. Explain the roles of modality-gated ion channels and generator potentials in the transmission of sensory information: _____

IV. Neurotransmitters (373)

1. What are neurotransmitters? _____

2. Name the four classes of known neurotransmitters.

    a. _____

    b. _____

    c. _____

    d. _____

3. What determines whether a given neurotransmitter will have an excitatory or inhibitory effect on a postsynaptic cell? _____

4. Compare action potentials and postsynaptic potentials. _____

5. Complete the following table.

Group	Probable Functions
Neuroactive peptides	
Biogenic amines (monoamines)	
Compound	
GABA, glycine, and glutamate	
	Excitatory inhibitory neuromuscular transmission Neuroglandular transmission involved in memory

6. Indicate the major "comment" that applies to each of the foregoing compounds that you have described in question 5.

a. _____

b. _____

c. _____

d. _____

e. _____

7. Explain the presumed roles of nitric oxide and carbon monoxide in long-term memory.

_____

8. What are the three fastest neurotransmitters?

a. _____

b. _____

c. _____

V. Neuronal Circuits (376)

Match the following pairs.

_____ 1. Negative feedback loop
_____ 2. Knee-jerk reaction
_____ 3. Open circuit
_____ 4. Neuron branches out to affect multiple postsynaptic neurons
_____ 5. Same information passed along different pathways
_____ 6. Receptor of postsynaptic neuron associated with multiple presynaptic neurons

a. divergence
b. convergence
c. feedback circuit
d. parallel circuits
e. two-neuron circuit
f. three-neuron circuit

VI. The Effects of Aging on the Nervous System (377)

1. Excluding a specific disease such as Alzheimer's or Parkinson's, what effects does age have on a person's mental capacity? _____

VII. Developmental Patterns of Neural Tissue (377)

1. What is meant by the term plasticity as it relates to neural tissue? _____

2. What is the benefit, if any, of plasticity? _____

VIII.   When Things Go Wrong (378)

1. Complete the following table.

Disease	Cause(s)	Symptom(s)
Huntington's disease		
Peripheral neuritis		
Parkinson's disease		
Lou Gehrig's disease		
Myasthenia gravis		
Multiple sclerosis		

2. What are seven ways in which drugs can affect the events that take place at synapses?

a. _____

b. _____

c. _____

d. _____

e. _____

f. _____

g. _____

Key Terms

acetylcholine 373
action potential 365
afferent (sensory) neurons 360
all-or-none principle 366
axon 357
axonal transport 357
axoplasmic flow 357
cell body 353
central nervous system (CNS) 353
chemically gated channels 364
conductivity 353
dendrite 357
depolarization 365
efferent (motor) neurons 360
end bulb 357
excitability 353

gated channel 364
interneuron 360
modality-gated channels 365
myelin 357
nerve 359
nerve impulse 365
neuroglia 361
neuron 353
neuronal circuit 376
neurotransmitter 373
node of Ranvier 357
open ion channels 364
peripheral nervous system 353
polarization 363
postsynaptic neuron 369
potential difference 363

presynaptic neuron 369
refractory period 366
repolarization 365
resting membrane potential 363
saltatory conduction 369
satellite cell 361
Schwann cell 357, 361
sodium-potassium pump 364
somatic nervous system 353
stimuli 353
supporting cells 353
synapse 357, 369
threshold stimulus 365
transmitter-gated channels 364
visceral nervous system 353
voltage-gated channels 364

Post
Test

Multiple Choice

___ 1. The central nervous system (CNS)
consists of the

    a. brain, spinal cord, and peripheral nerves
    b. brain and cranial nerves
    c. brain only
    d. brain and spinal cord only
    e. spinal cord and peripheral nerves only

___ 2. The peripheral nervous system
(PNS) consists of the

    a. neuronal processes emerging from and
       going to the brain and spinal cord
    b. all nerves and ganglia located outside the
       CNS
    c. somatic and visceral nervous systems
    d. somatic motor, somatic sensory, visceral
       motor, and visceral sensory divisions of
       the PNS
    e. all of the above

___ 3. The cranial nerves are represented as
part of the

    a. PNS
    b. CNS
    c. autonomic nervous system
    d. visceral motor system
    e. both c and d

___ 4. Most axons convey nerve impulses
(action potentials)

    a. toward the cell body
    b. away from the cell body
    c. toward the dendrites of the same cell
    d. away from the telodendria
    e. both c and d

___ 5. Some axons are covered with a
lipid-rich sheath called

    a. Schwann cells
    b. neurilemma
    c. a medullary sheath
    d. a myelin sheath
    e. oligodendrocytes

Completion

6. The covering of myelinated fibers is interrupted at regular intervals by gaps called
_____.

7. In the PNS this sheath is formed by _____cells.

8. The sheaths of CNS axons are formed by _____.

9. A collection of neuron cell bodies belonging to the PNS is known as a _____.

10. These cell bodies are surrounded by _____, which are the equivalents of
Schwann cells.

Identify the structures indicated in the figure below.
___ 11. axon
___ 12. node of Ranvier
___ 13. layers of myelin
___ 14. oligodendrocyte
___ 15. Schwann cell

Matching

___ 16. Intraneuron transport away from cell body only
___ 17. Sensory neurons of PNS
___ 18. Convey signals over long distances
___ 19. Differentiate into neurons
___ 20. Two-way intraneuron flow
___ 21. Carry action potentials from CNS to effector organs
___ 22. Most common sensory neurons in PNS
___ 23. The "trigger" zone of a neuron
___ 24. -70 millivolts
___ 25. Action potential jumps from node to node (of Ranvier)

a. neuroblasts
b. resting membrane potential
c. afferent
d. axonal transport
e. saltatory conduction
f. initial segment
g. efferent
h. relay interneurons
i. axoplasmic flow
j. unipolar

Completion

26. A self-regulating transport system located within the plasma membrane that maintains a constant electrical differential between the inside and outside of the cell is the
_____.

27. This system works by selectively pumping positively charged ions of one kind out of the cell and positively charged ions of another kind into the cell in the ratio of about 3 to 2, respectively. These ions are _____ and _____.

28. Since more positive ions (cations) are pumped out than in, the inside of the cell becomes relatively _____ to the outside.

29. The differential charge that is thus maintained across the plasma membrane is known as a
_____.

30. The energy that drives this cation pump is derived from the hydrolysis of _____.

Matching

___ 31. A membrane channel that is open to facilitate the maintenance of a resting membrane potential

___ 32. Activated to open when the membrane potential changes

___ 33. Impermeable to large protein molecules

___ 34. Responds to specific conditions, such as temperature

___ 35. Activated by neurotransmitter molecules

a. voltage-gated channel
b. chemically gated channel
c. modality-gated channel
d. open channel

___ 36. A change external to a neuron that might affect the resting membrane potential of that neuron

___ 37. A change of a magnitude sufficient to initiate a nerve impulse

___ 38. A nerve impulse

___ 39. The progression of an impulse along a myelinated neuronal fiber

___ 40. The time immediately following the generation of a nerve impulse during which another threshold stimulus will fail to elicit another impulse

___ 41. The restoration of a resting membrane potential

___ 42. A neuron that can, by virtue of its position, pass on an action potential to another neuron

___ 43. − 70 mV

a. action potential
b. refractory period
c. threshold stimulus
d. stimulus
e. resting membrane potential
f. repolarization
g. presynaptic neuron
h. saltatory conduction

___ 44. Space between pre- and postsynaptic neurons

___ 45. Uses connexons

___ 46. Functional conjunction between axons of different neurons

___ 47. Functional conjunction between axon of one neuron and cell body of another neuron

___ 48. Functional conjunction between axon of one neuron and dendrite of another

___ 49. Most use acetylcholine or noradrenaline

___ 50. Joined by gap junction

___ 51. Speediest conduction rate

___ 52. Site of neurotransmitter hydrolysis

a. axodendritic synapse
b. axoaxonic synapse
c. axosomatic synapse
d. electrical synapse
e. chemical synapse
f. synaptic cleft

Multiple Choice

___ 53. Two basic types of responses that can be produced at different synaptic sites on the postsynaptic membrane

    a. PSP and PIP
    b. EPSP and EPIP
    c. EPSP and IPSP
    d. ESP and SSP

___ 54. The excitatory one, although it fades out, makes the postsynaptic membrane more excitable, an effect called

    a. facilitation
    b. gradation
    c. inhibition
    d. hyperpolarization

___ 55. When K^+ and Cl^- channels open and K^+ rushes out and Cl^- rushes in through the postsynaptic membrane, the result is the creation of

    a. EPSP
    b. APSP
    c. ASAP
    d. IPSP

___ 56. The additive effect of the synaptic inputs from different neurons on one receptor neuron is called

    a. temporal summation
    b. spatial summation
    c. facilitation
    d. sensitization

___ 57. Inhibition is

    a. ineffective in most cases
    b. a last resort when neurons fail to fire
    c. as important as excitation in many bodily functions
    d. rarely used in normal activities of the body

Matching

___ 58. Excitatory and inhibitory; neuromuscular and/or neuroglandular transmission; involved in memory

___ 59. Inhibits secretion of growth hormone

___ 60. Mainly inhibitory, excitatory in brain; important in the spinal cord.

___ 61. Involved in sleep, mood, appetite, and pain

___ 62. Inhibitory; evokes IPSPs in brain neurons

___ 63. Implicated in neurodegenerative diseases, such as Altzheimer's disease; it has a variety of functions

    a. acetylcholine
    b. GABA
    c. glycine
    d. dopamine
    e. norepinephrine
    f. serotonin
    g. somatostatin
    h. endorphins
    i. nitric oxide
    j. substance P

Label the Diagram

___ 64. oligodendrocyte
___ 65. node of Ranvier
___ 66. microglial cell
___ 67. astrocyte
___ 68. myelin sheath
___ 69. neuron
___ 70. ependyma

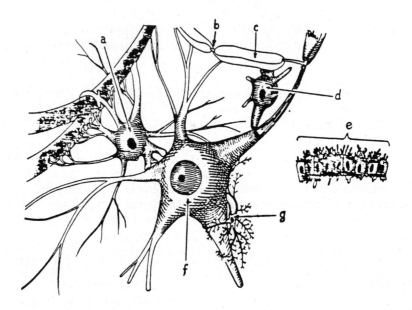

Integrative Thinking

1. Normal cells divide at a rate that is commensurate with the need for repair or replacement. Tumors result when this rate is not longer controlled, and the cells divide at an unregulated rate, forming masses of tissue that displace normal cells and, in due course, may become invasive and life-threatening. The brain is not exempt from tumors. Explain how this can be, in view of the fact that the neurons of most living mammals, including human beings, do not and cannot divide.

2. Diuretics increase urine production as treatment for symptoms of excess water retention. Individuals taking diuretics are cautioned to get plenty of potassium in their diet in order to compensate for its excessive loss in high urine output. Explain what effects reduced quantities of potassium in the body fluids would have on the nervous system. Be specific in your answer.

3. Curare is a well-known drug that blocks the acetylcholine receptors at the motor end plate. It is frequently used by surgeons as an aid in certain kinds of operations. Why? (Understand, of course, that its effects can be neutralized at any time.)

Your
Turn

Write each of the topics in the list below on a 3 x 5 card. Put the cards in a hat, and let each member of your team, with eyes closed, select a card. Taking turns, let each card-carrying member announce what is on the card and proceed to explain the similarities and differences of the topic as it applies to skeletal muscle fibers vs. neurons. Following each turn, let the rest of the team members add to, detract from, or make clarifying comments about the answers given. By sharing in this way you will all benefit. Please let your instructor know how this experimental game for understanding works out.

List of Topics

1. Graded responses
2. All-or-none
3. Roles of calcium
4. How impulses travel through or along the cell
5. Synapse structure
6. Response to stimulation
7. Structure of cells

13 The Central Nervous System I: The Brain

Active Reading

Introduction

1. How much does the brain weigh? _____

2. How many neurons might there be in the human brain? _____
   How many synapses? _____

I. General Structure of the Brain (385)

1. What is the technical term for the brain? _____

2. Complete the table.

Structure	Description	Major Function
brainstem		
midbrain		
pons		
medulla oblongata		
	second largest part of brain	
	deep portion of brain	
		center for all conscious living
	outer layer of cerebrum	
basal nuclei		
		connects cerebral hemispheres

3. By what protective structures is the brain surrounded? _____

4. Box 13.1 describes a protective mechanism of extreme importance. What is it, and under what circumstances may it be too effective? _____

_____

II. Meninges, Ventricles, and Cerebrospinal Fluid (385)

1. Why is there no epidural space in the cranial cavity? _____

2. Contrast the structures of the inner and outer layers of the dura mater. _____

_____

3. What lies beneath the dura mater, separating it from the arachnoid? _____

4. What kind of tissue composes the arachnoid? _____

5. What lies beneath the arachnoid, separating it from the pia mater? _____

6. In which layer of the meninges lie the blood vessels that supply blood to the brain?_____

_____

7. List in order, from top to bottom, the ventricles of the brain.

   a. _____

   b. _____

   c. _____

8. Through what space are the lateral ventricles connected to the third ventricle?_____

9. Through what space are the third and fourth ventricles connected? _____

10. What kind of fluid is the cerebrospinal fluid (CSF)? _____

11. What physical functions does the CSF serve? _____

12. What chemical functions does the CSF serve? _____

13. Where is the CSF formed? _____. How is it formed? _____

_____

14. Starting at the superior-most area, in what order does the CSF pass through the areas listed below? ___, ___, ___, ___, ___, ___, ___

   a. cerebral aqueduct

   b. cisterna magna

   c. two lateral ventricles

   d. intraventricular foramina

   e. apertures in roof of fourth ventricle

   f. third ventricle

   g. fourth ventricle

15. How do the arachnoid villi regulate the pressure of the cerebrospinal fluid in the subarachnoidal space? _____

III. Nutrition of the Brain (391)

1. How much blood reaches the brain each minute? _____

___ 2. What percentage of the body's need
for oxygen does the brain account
for?

    a. 5%
    b. 10%
    c. 20%
    d. 40%
    e. 0% (the brain functions anaerobically
       only)

3. What is hypoxemia? _____

4. What is hyperoxemia? _____. What are its symptoms? _____

_____

5. Returning now to the blood-brain barrier, describe how it regulates the transfer of substances across it? _____

6. To what kinds of substances is the blood-brain barrier passively permeable? _____

7. What role do astrocytes play in the barrier? _____

IV. Brainstem (392)

1. What is the basic function of the brainstem? _____

2. What are the three segments of the brainstem?

    a. _____

    b. _____

    c. _____

3. Name five other structures of the brainstem.

    a. _____

    b. _____

    c. _____

    d. _____

    e. _____

Match the following pairs:

___ 4. Bundle of axons
___ 5. Sequence of motor neurons going from higher
to lower areas of the CNS
___ 6. Collection of neuron cell bodies inside central
nervous system
___ 7. Collection of neuron bodies outside central
nervous system
___ 8. Sequence of nuclei and tracts
___ 9. Sequence of sensory neurons extending
upward from spinal cord

    a. nucleus
    b. ganglion
    c. tract
    d. pathway
    e. ascending pathway
    f. descending pathway

10. In what parts of the brainstem is the reticular formation found?

    a. _____

    b. _____

    c. _____

____ 11. Which of the following functions does the reticular formation serve?

    a. influences brain's state of arousal
    b. regulates breathing, heart rate, and blood flow
    c. influences dull and diffuse sensation of pain
    d. engenders feelings of fear and terror
    e. all of the above
    f. three of the above

12. List seven functions of the reticular activating system.

    a. _____

    b. _____

    c. _____

    d. _____

    e. _____

    f. _____

    g. _____

13. Where in the cranial cavity does the medulla oblongata lie? _____

14. What vital functions are associated with the reticular formation component of the medulla oblongata? _____

15. What is meant by the term decussation? _____

16. What is the significance of the decussation of the corticospinal tracts? _____

17. What regions of the brain do the inferior cerebellar peduncles connect?

    a. _____

    b. _____

18. What does the pons form? _____

Match the following.

____ 19. Pontine nuclei
____ 20. Corticopontocerebellar pathway
____ 21. Senses of touch-pressure and proprioception
____ 22. Rubrospinal tract
____ 23. Senses of pain, temperature, and light touch
____ 24. Communication between cerebral cortex and the cerebellum of the opposite side

    a. dorsal pons
    b. ventral pons

25. What functions are associated with the inferior colliculi of the midbrain? _____

_____

26. Describe the location and function of the following.

    a. cerebral peduncles _____

    b. corpora quadrigemina _____

    c. nucleus ruber _____

    d. substantia nigra _____

V. Cerebellum (399)

1. What is the main role of the cerebellum? _____

2. What separates the cerebellum from the occipital lobes of the cerebrum? _____

3. What are the three parts of the cerebellum?

    a. _____

    b. _____

    c. _____

4. On the figure below, locate the following: occipital lobe of cerebrum, cerebellum, folia cerebelli, arbor vitae, cerebellar cortex, midbrain, pons, fourth ventricle, medulla, spinal cord.

5. What movements does the cerebellum initiate? _____

6. What are proprioceptive inputs? _____

7. Where do the cerebellar pathways decussate?

    a. _____

    b. _____

VI. Cerebrum (399)

1. How many neurons does the cerebral cortex contain? _____

2. Describe the form and function of the following.

    a. gyri _____

    b. sulci _____

    c. fissures _____

3. What part of the cerebrum is gray matter? _____. Where is white matter of the cerebrum? _____

4. Of what is white matter composed? _____

5. Describe the functions of each of the following

    a. projection fibers _____

    b. association fibers _____

    c. commissural fibers _____

    d. basal nuclei _____

6. What functions are associated with each hemisphere of the crebral cortex?

    a. left _____

    b. right _____

____ 7. Who would more likely be left
        handed?

    a. mathematician
    b. sculptor
    c. forester
    d. botanist

Match the following pairs.

____ 8. Sylvius
____ 9. Rolando

    a. central sulcus
    b. lateral cerebral sulcus

10. What two gyri contribute to the limbic lobe?

    a. _____

    b. _____

Match the following pairs.

____ 11. Functions in hearing, equilibrium, and to a certain
         degree, emotion and memory
____ 12. Motor control of voluntary movements, emotional
         expressions, morals, and ethics
____ 13. Perception of odors, hallucinations of odors, fear,
         and unreality
____ 14. An island of gastrointestinal and other visceral
         activities
____ 15. Primary visual cortex, understanding of the visual
         and spoken word
____ 16. An awareness of the body in relation to its external
         environment, general senses, and taste

    a. frontal lobe
    b. parietal lobe
    c. temporal lobe
    d. occipital lobe
    e. central lobe
    f. limbic lobe

17. What are the components of the limbic system? _____

_____

_____

VII. Diencephalon (409)

1. Label the diagram below, using the letters for reference.

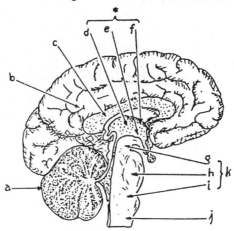

a. _____

b. _____

c. _____

d. _____

e. _____

f. _____

g. _____

h. _____

i. _____

j. _____

k. _____

* _____

2. In addition to housing the third ventricle, of what parts is the diencephalon composed?

a. _____

b. _____

c. _____

d. _____

3. Where is the diencephalon located? _____

4. Describe the four major areas of activity associated with the thalamus.

a. _____

b. _____

c. _____

d. _____

5. What does the word "hypothalamus" mean? _____

6. Describe the eight important functions of the hypothalamus.

   a. _____

   b. _____

   c. _____

   d. _____

   e. _____

   f. _____

   g. _____

   h. _____

7. Where is the epithalamus located? _____

8. What is the function of the ventral thalamus? _____

VIII. Regeneration in the Central Nervous System (411)

1. List four reasons for the fact that severed axons of the mature nervous system may be unable to regenerate.

   a. _____

   b. _____

   c. _____

   d. _____

IX. Learning and Memory (411)

1. In what specific area of the brain does learning take place? _____

2. By what age does the dominance of one cerebral hemisphere over the other become fairly well fixed? _____

3. Describe the following terms.

   a. memory traces _____

   b. short-term memory _____

   c. long-term memory _____

4. What two gases may have an important role in the transfer of thoughts from short-term memory to long-term memory?

   a. _____

   b. _____

5. In what areas of the brain are short-term memories converted into long-term? _____

X. Sleep and Dreams (412)

1. For what do the initials EEG stand? _____

2. Complete the following table.

Wave Pattern	State of Mind
Theta waves	
	Alert, stimulated
Length of wave	
8 to 12 cps	

3. Name and describe the four stages of sleep.

   a. _____

   b. _____

   c. _____

   d. _____

4. What do we experience during REM sleep? _____

5. What is the importance of REM sleep? _____

XI. Sleep Patterns and Age (414)

1. What may account for reduced short-term memory among the elderly? _____

XII. Chemicals and the Nervous System (415)

1. Complete the following table.

Class of Drug	Mechanism of Action	Example
Stimulants	Block uptake of norepinephrine and serotonin at nerve endings	
		Aspirin
	Inhibit activity of central nervous system	
Antianxiety		
		Mescaline

2. What differences in drug absorption and clearing are exhibited by the elderly? _____

_____

XIII. Developmental Anatomy of the Brain (417)

1. Describe the development of the brain from 8 days after conception to the fourth fetal week. _____

_____

Match the following pairs.

____ 2. Mesencephalon   a. forebrain
____ 3. Prosencephalon   b. midbrain
____ 4. Rhombencephalon   c. hindbrain

5. Into what do the following parts divide during the seventh week?

     a. prosencephalon _____

     b. rhombencephalon _____

6. When do the sulci begin to develop? _____

7. What does the brain weigh at birth? _____

XIV. When Things Go Wrong (420)

1. What are the causes of the following maladies?

     a. senile dementia _____

     b. cerebral palsy _____

     c. stroke (CVA) _____

     d. dyslexia _____

     e. encephalitis _____

     f. migraine headaches _____

2. Describe possible treatment for Parkinson's disease. _____

3. Distinguish grand mal epilepsy and petit mal epilepsy. _____

4. What symptoms result from lead poisoning? _____

_____

Key Terms

basal nuclei 385, 402
blood-brain barrier 396
brainstem 385, 392
cerebellum 385, 399
cerebral lobes 404
cerebrospinal fluid 390
cerebrum 385, 399
corpus callosum 385, 401

diencephalon 385, 409
epithalamus 410
fissure 401
gyri 401
hypothalamus 409
limbic system 408
medulla oblongata 393
meninges 385

midbrain 398
pons 398
REM sleep 414
reticular formation 393
sulci 401
thalamus 409
ventral thalamus 410
ventricle 387

Post Test

Matching

____ 1. main function: muscular coordination
____ 2. corpus callosum
____ 3. medulla oblongata
____ 4. hypothalamus
____ 5. basal nuclei
____ 6. pons
____ 7. cortex
____ 8. thalamus
____ 9. reticular formation
____ 10. lateral ventricles

a. brainstem
b. cerebellum
c. diencephalon
d. cerebrum

____ 11. cerebral peduncles
____ 12. substantia nigra
____ 13. vasomotor and respiratory centers
____ 14. pneumotaxic center
____ 15. arbor vitae
____ 16. corpora quadrigemina
____ 17. tubercula gracilis and cuneatus
____ 18. pyramids
____ 19. vermis
____ 20. red nucleus

a. medulla
b. pons
c. midbrain
d. cerebellum

____ 21. Broca's speech area
____ 22. auditory cortex
____ 23. primary somesthetic association area
____ 24. primary visual cortex
____ 25. the "motor lobe"
____ 26. involved in visual pursuit
____ 27. Wernicke's area
____ 28. control of moral and ethical behavior
____ 29. closest to ears: critical role in hearing and equilibrium
____ 30. postcentral and supramarginal gyri

a. frontal lobes
b. parietal lobes
c. temporal lobes
d. occipital lobes

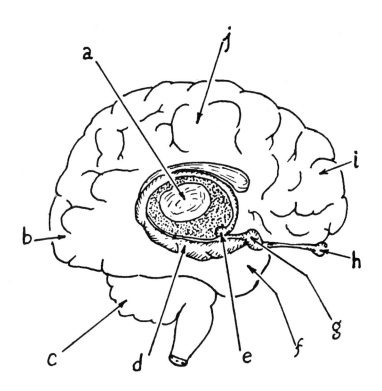

Label the figure above (use letters for reference).

____ 31. amygdala
____ 32. hippocampus
____ 33. frontal lobe
____ 34. occipital lobe
____ 35. thalamus
____ 36. olfactory bulb
____ 37. mamillary body
____ 38. cerebellum
____ 39. parietal lobe
____ 40. temporal lobe

Multiple Choice

____ 41. A slender network of neurons and fibers, located deep in the brainstem

    a. globus pallidus
    b. flocculonodular lobes
    c. reticular formation
    d. tectum
    e. inferior colliculi

____ 42. Ascending pathways to the cerebrum form the reticular activating system, which influences the brain's state of

    a. wakefulness
    b. quiescence
    c. coordinate behavior
    d. initiative
    e. somnolence

_____ 43. The corticospinal (pyramidal) tracts
of the medulla oblongata

    a. have purely sensory functions
    b. decussate
    c. form peduncles
    d. form the ventral pons
    e. form the dorsal pons

_____ 44. The part of the brain that is especially involved with coordinating
agonists and antagonists in skeletal
muscle

    a. cerebrum
    b. diencephalon
    c. limbic system
    d. pons
    e. cerebellum

_____ 45. The part of the brain that continu-
ously monitors sensory input from
muscles, joints, tendons, and organs
of equilibrium

    a. cerebrum
    b. diencephalon
    c. limbic system
    d. pons
    e. cerebellum

_____ 46. The occipital lobes function mainly
for processing

    a. proprioceptor input
    b. auditory input
    c. visual input
    d. olfactory input
    e. tangoreceptor input

_____ 47. An assemblage of certain cerebral,
diencephalic, and midbrain structures that is actively involved in
memory and emotions and the
visceral and behavioral responses
that are associated with them

    a. limbic system
    b. epithalamus
    c. hypothalamus
    d. thalamus
    e. corpus callosum

_____ 48. Located in the center of the cranial
cavity and forming the lateral walls
if the third ventricle

    a. pituitary gland
    b. amygdala
    c. mamillary body
    d. cerebellar peduncles
    e. thalamus

_____ 49. The part of the brain that functions
in integrating via the autonomic
nervous system such essential
"house-keeping" activities as temperature regulation, blood pressure,
water and electrolyte balance, sleep/
wake patterns, sexual responses, and
many others

    a. limbic system
    b. epithalamus
    c. hypothalamus
    d. thalamus
    e. corpus callosum

_____ 50. The structure that produces oxytocin
and antidiuretic hormone, as well as
certain releasing hormones

    a. limbic system
    b. epithalamus
    c. hypothalamus
    d. thalamus
    e. ventral thalamus

Label the diagram

____ 51. precentral gyrus
____ 52. postcentral gyrus
____ 53. longitudinal fissure
____ 54. central sulcus
____ 55. fissure of Rolando

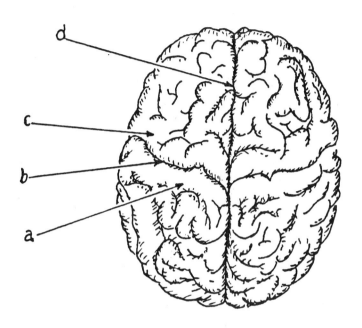

Matching

____ 56. Massive bundles of axons connecting left and right hemispheres
____ 57. Caudate and putamen
____ 58. Axons that connect cerebral cortex to other brain structures, such as thalamus and brainstem
____ 59. Link one area of the cortex with another cortical area in the same hemisphere
____ 60. Malfunctions expressed as dyskinesias, as in Parkinson's disease and Huntington's disease

a. basal nuclei
b. association fibers
c. projection fibers
d. corpus callosum

____ 61. Under control of hypothalamus
____ 62. Neurotransmitters of a different sort
____ 63. Brain waves typical of an alert, stimulated brain
____ 64. Brain waves typical of deep sleep, damaged brain, or of infants
____ 65. The time for dreams

a. beta
b. delta
c. NO and CO
d. REM
e. ADH

Matching

____ 66. MAOIs, Ritalin
____ 67. Ether, chloroform, ethyl alcohol
____ 68. Barbituates, Seconal, phenobarbitol
____ 69. Cannabis, hashish
____ 70. Caffeine, nicotine
____ 71. Meprobamate (Miltown), Vallium
____ 72. Lysergic acid diethylamine
____ 73. Amphetamines, Dexedrine
____ 74. Prozac, Parnate
____ 75. Cocaine, cola drinks

a. stimulant
b. depressant
c. antidepressant
d. psychedelic/hallucinogenic

Integrative Thinking

1. Ms. Goldsmith was worried about her seventeen-year-old daughter, Margaret. Margaret had changed since she had made certain new friends at school. She was "itchy," high-strung, and irritable. Margaret never had time for breakfast anymore, or lunch, for that matter. And when her mother told her she was getting awfully thin, she blew up at her and as much as told her to mind her own business. That wasn't her Margaret! No way! So Ms. Goldsmith called her doctor and asked her to see her daughter. The doctor found that Margaret had a rapid pulse rate, dilated pupils, and elevated blood glucose. What do you think was her diagnosis? Why? What action would you recommend to Ms. Goldsmith?

2. A driver who had neither an air bag nor a functional seat belt found himself sprawled across his steering wheel with his head down, looking at the windshield wipers. Remarkably, he recovered from the cuts, bruises, and concussion sustained in the accident, except for a mild but permanent ataxia of the right arm and hand. Explain.

3. The case of the phantom limb: Persons who have lost an arm or a leg feel a muscle cramp or, more frequently, a burning pain in that arm or leg as if it were still intact. Proposals that the sensations are due to spontaneous firing of the stumps of sensory nerves that supplied the erstwhile limb have proved untenable, as have any happenings in the spinal cord. Where would you expect the sensations to originate, and what other areas would you expect to be involved in the case of the phantom limb?

Your Turn

How about drawing some more cards from a hat? This time, on each card write the name of a condition or abnormality, such as the ones suggested below. Then put them in the hat, shake them up, and let each member of your study group draw one and identify the probable cause of the condition and/or the area of the brain that is involved. Following each person's answer, the group can clarify, correct, or add to the answer.

Suggested entries:

inability to articulate spoken words, inability to say words that make sense, loss of feeling in some particular part of the body, polyurea (excessive urine production), memory, inability to track a moving object visually, insomnia. Surely you can think of more.

14 The Central Nervous System II: The Spinal Cord

Active Reading

Introduction

1. How many pairs of spinal nerves connect with the spinal cord? _____

2. What two important functions does the spinal cord serve?

    a. _____

    b. _____

3. What do somatic spinal reflexes involve? _____

4. What are the functions of visceral spinal reflexes? _____

I. Basic Anatomy of the Spinal Cord (429)

1. Between what two locations does the spinal cord extend?

    a. _____

    b. _____

2. Of what kind of tissue is the filum terminale composed? _____

3. What is the conus terminalis? _____

4. At what vertebral level are the cervical and lumbosacral enlargements?

    a. cervical _____

    b. lumbosacral _____

5. Complete the following table.

Region	Number of Nerve Pairs
Cervical (C)	
Thoracic (T)	
Lumbar (L)	
Sacral (S)	
Coccygeal (Co)	

6. What are the lumbar and sacral roots collectively called? _____

7. In the figure below, identify the following meninges and spaces: dura mater, arachnoid, pia mater, epidural space, subdural space, subarachnoid space.

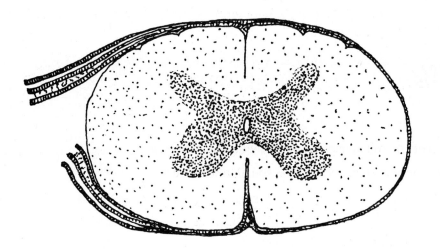

8. Where does the cerebrospinal fluid (CSF) enter the spinal cord? _____

9. Name two functions of the CSF.

    a. _____

    b. _____

10. What is a spinal tap, and how is it performed? _____

11. What may a marked increase in white blood cells in the CSF indicate? _____

12. In the figure below, identify the following: anterior median fissure, posterior median sulcus, gray matter, posterior horns, anterior horns, gray commissure, central canal, white matter, anterior funiculus, lateral funiculus, posterior funiculus.

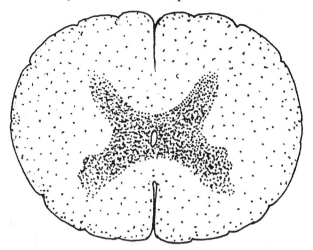

13. Of what types of cells is the gray matter composed? _____

14. What gives white matter its color? _____

15. What is the function of the ascending tracts? _____

16. What is the function of the descending tracts? _____

17. Distinguish between the functions of long tracts and short tracts. _____

_____

Match the following.

____ 18. Anterior	a. ventral roots
____ 19. Posterior	b. dorsal roots
____ 20. Sensory	
____ 21. Motor	
____ 22. Afferent	
____ 23. Generally efferent	

II. Functional Roles of Pathways of the Central Nervous System (436)

1. What are processing centers, and what do they do? _____

2. Where do the upper motor neurons originate? _____

Match the following.

____ 3. Innervate extrafusal muscle fibers	a. upper motor neurons
____ 4. Innervate intrafusal muscle fibers	b. lower motor neurons
____ 5. Terminate outside of neuromuscular spindle	c. alpha motor neurons
	d. gamma motor neurons
____ 6. Form voluntary descending pathways	
____ 7. Comprise pyramidal and corticobulbar tracts	
____ 8. Entire neurons within central nervous system	
____ 9. Influence smoothness of stretch reflex reaction	

10. Name and describe the point(s) of origin of four upper motor neuron tracts.

   a. _____

   b. _____

   c. _____

   d. _____

11. What three parts make up each processing center?

   a. _____

   b. _____

   c. _____

Match the following pairs:

____ 12. Extends from spinal cord or brainstem to a   a. first-order neuron
         nucleus in the thalamus   b. second-order neuron
____ 13. Extends from the thalamus to a sensory area   c. third-order neuron
         of the cerebral cortex
____ 14. Extends from the sensory receptor to the
         central nervous system

15. What is the implication of the crossing over of sensory pathways? _____

16. What does the anterolateral system consist of? _____

17. Describe the structure and function of the posterior column medial lemniscus pathway.

   _____

III. Spinal Reflexes (439)

1. What is a reflex? _____

2. What is the difference between somatic and visceral reflexes? _____

Match the following.

____ 3. Direct synapse between sensory   a. polysynaptic reflex arc
        neuron and motor neuron   b. monosynaptic reflex arc
____ 4. One or more interneurons synapse
        with motor and sensory neurons
____ 5. Myotactic reflex
____ 6. Pulling foot up and away from a
        sharp tack

7. Define ipsilateral. _____

8. What is the function of the gamma motor neuron reflex arc? _____

9. What is the role of flexor reflex afferents? _____

10. What is the role of commissural interneurons in a crossed-extensor reflex? _____

    _____

11. What is the significance of Babinski's reflex? _____

Label the parts of the diagram below as indicated.

___ 12. dorsal root
___ 13. axon of sensory neuron
___ 14. ventral root
___ 15. axon of motor neuron
___ 16. interneuron
___ 17. synapses
___ 18. posterior (dorsal) ramus
___ 19. sensory neuron cell body
___ 20. posterior (sensory) root ganglion
___ 21. anterior (ventral) ramus
___ 22. motor neuron cell body

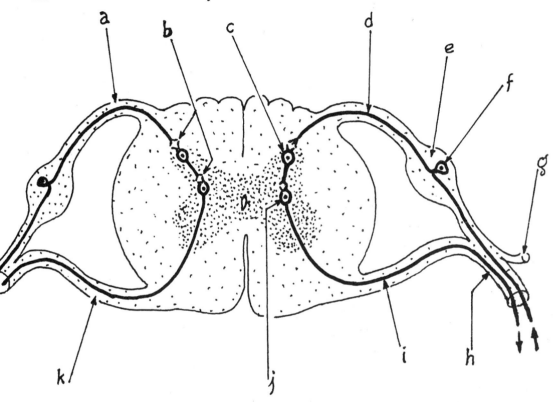

23. Define the following terms.

    a. tonus _____

    b. hypotonia _____

    c. hypertonia _____

    d. atony _____

    e. hyporeflexia _____

24. What are the clinical significances of hyperreflexia and hypertonia? _____

25. Briefly describe and indicate the clinical significance of each of the following reflexes.

Reflex	Description	Indication
Biceps		
Kernig's		
Triceps		
Plantar (achilles)		

IV. Developmental Anatomy of the Spinal Cord (444)

1. What initiates the embryonic development of the nervous system? _____

2. On the figures below, identify the following: ectoderm, endoderm, neural crest, neural plate, neural tube, neural groove, matrix layer, mantle layer, marginal layer. Additionally, indicate the approximate stage of development each of the figures represents.

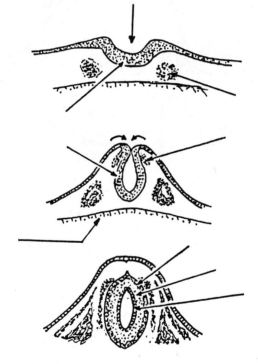

V. When Things Go Wrong (445)

1. Describe the causes and symptoms of the following.

    a. paraplegia _____

    b. quadriplegia _____

    c. hemiplegia _____

2. What events take place immediately after a spinal cord injury? _____

3. Describe five new treatments for spinal cord injuries.

   a. _____

   b. _____

   c. _____

   d. _____

   e. _____

4. Complete the following table.

Disease	Cause(s)	Symptom(s)	Treatment
Carpal tunnel syndrome			
Poliomyelitis			
Sciatica			
Shingles, or herpes zoster			
Spinal meningitis			

Key Terms

anterior horns 432
anterolateral system 438
ascending tracts 432
cauda equina 429
descending tracts 432
dorsal root 432
dorsal root ganglia 432
funiculi 432

gamma motor neuron reflex 440
gray commissure 432
gray matter 431
lower motor neurons 436
meninges 429
posterior column-medial lemniscus pathway 438
posterior horns 432

reflex 439
reflex arc 439
stretch reflex 440
upper motor neurons 437
ventral root 432
white matter 432

Post Test

Multiple Choice

_____ 1. The spinal meninges are the same as and are continuous with

a. all the meninges of the brain
b. the filum terminale
c. the denticulate ligament
d. the cauda equina
e. none of the above

_____ 2. The spinal cord is partially divided into left and right halves by the

a. anterior and posterior horns
b. lateral horns
c. gray commissure
d. anterior and posterior median fissures or sulci
e. three pairs of funiculi

_____ 3. White matter of the spinal cord is concentrated in

a. ascending tracts
b. descending tracts
c. anterior horns
d. posterior and lateral horns
e. both a and b

_____ 4. The neurological function of the descending tracts is

a. maintenance of both motor and sensory activities
b. maintenance of involuntary activities only
c. transmission of action potentials of sensory neurons only
d. transmission of action potentials of efferents only
e. all of the above

_____ 5. With exceedingly few, if any, exceptions

a. dorsal roots are motor and ventral roots are sensory
b. ventral roots are motor and dorsal roots are sensory
c. both dorsal and ventral roots are mixed
d. spinal ganglia contain the cell bodies of motor neurons
e. ventral rami contain only efferent axons

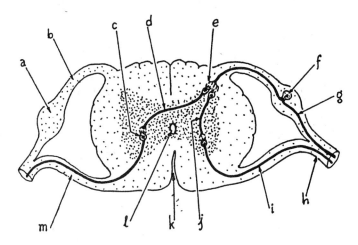

Identify each of the structures indicated in the diagram, using the letters for reference.
___ 6. spinal nerve
___ 7. ipsilateral interneuron
___ 8. commissural interneuron
___ 9. central canal
___ 10. posterior root ganglion
___ 11. axon of afferent neuron
___ 12. axon of efferent neuron
___ 13. anterior median fissure
___ 14. cell body of a motor neuron
___ 15. cell body of a unipolar neuron
___ 16. posterior root
___ 17. anterior root
___ 18. synaptic junction

Matching

___ 19. Extend from sensory receptors to CNS
___ 20. Include the rubrospinal tract
___ 21. Sensory neurons entirely within the brain
___ 22. Innervate intrafusal muscle fibers
___ 23. Innervate extrafusal fibers
___ 24. Convey action potentials from spinal cord to thalamus
___ 25. Tracts originating from nuclei in lower brainstem

a. alpha motor neurons
b. gamma motor neurons
c. upper motor neurons
d. first-order neurons
e. second-order neurons
f. third-order neurons

Matching Alternatives

___ 26. A reflex involving skeletal muscle is (a) a somatic reflex, or (b) a visceral reflex.
___ 27. Reflexes involving involuntary homeostatic functions are (a) somatic, or (b) visceral.
___ 28. Reflexes that are carried out by neurons in the spinal cord alone constitute (a) spinal reflexes, or (b) subcerebral reflexes.
___ 29. In most polysynaptic reflexes (a) only agonist muscles react, or (b) agonists contract and antagonists are inhibited.
___ 30. Withdrawal reflexes involve (a) ipsilateral arcs only, or (b) both ipsilateral and contralateral arcs.

Matching

____ 31. When positive, meningeal irritation
        or herniated disk is indicated

____ 32. When negative, indicative of lesions
        of peripheral nerves or lower part of
        spinal cord; also of MS

____ 33. When positive in an adult, damage
        to upper motor neurons is indicated

____ 34. When negative, may indicate
        chronic diabetes or syphilis

____ 35. Indicative of dorsal column injury

a. abdominal reflex
b. Romberg's reflex
c. Babinski's reflex
d. Brudzinski's reflex
e. patellar reflex

____ 36. Includes the ependyma
____ 37. Includes mantle and marginal layers
____ 38. Releases a chemical releasing factor
____ 39. Parent to all spinal root ganglia,
        among other things
____ 40. Predecessor to the neural folds

a. notochord
b. neural crest
c. neural tube
d. neural plate

Integrative Thinking

1. One morning your ten-year-old child complains of a headache and nausea. You check his temperature and find he is running a slight fever, so you give him some aspirin and keep him in bed. He gets sicker, rather than better, and you become alarmed. His temperature flares up sharply, his neck is stiff, and he develops back spasms that cause his body to arch upward. You rush him to the hospital. What do you think the diagnosis will be? What other symptoms would you anticipate might develop to support this diagnosis? What tests might be made to confirm it? What evidence would indicate whether the causative agent is bacterial or viral?

2. In a withdrawal reflex, why are the antagonistic muscles inhibited? Why is a crossed extensor reflex usually superimposed on the ipsilateral reflex?

3. Suppose one has a lesion in the fasciculus gracilis in the posterior column of the spinal cord at the level of the fifth thoracic vertebra (T5). What sensory effect would this have, where would the effect be manifested, and on what side of the body?

Your Turn

1. Easy Does It! Taking turns, try out some of the reflexes that are described in Table 14.3 of your text, such as the Achilles, Babinski's, biceps, triceps, and patellar reflexes. Care must be taken to execute the necessary procedures gently, to avoid possible injuries, such as bruising.

2. Tract Meet. Prepare cards that have the name of each of the ascending and descending nerve tracts, one name per card. Shuffle them and put them in the old hat. Then, one by one, let each member of the study group, with eyes closed, draw a card from the hat, read out the tract named, and give its origin, course, and termination, and, as an extra challenge, the sensation (for sensory tracts) or motor impulse (for motor tracts) conveyed. The rest of the group should be prepared to aid, correct, or supplement the lucky card holder. The person or persons who give the most complete and correct answers should be given a special award by the rest of the group.

15 The Peripheral Nervous System

Active Reading

Introduction

1. What does the peripheral nervous system (PNS) consist of? _____

2. What is the difference between afferent and efferent neurons in respect to their relation to the CNS? _____

3. The PNS may be divided on a functional basis into the _____ nervous system and the _____ nervous system.

I. Classification of Functional Components of Nerves (451)

Match the following.

___ 1. General somatic afferent	a. from the central nervous system
___ 2. General visceral afferent	b. from the brain
___ 3. General somatic efferent	c. from the skin, skeletal muscles, joints, and connective tissue
___ 4. General visceral efferent	d. from the visceral organs
___ 5. Special visceral efferent	e. from the receptors of the olfactory, optic, auditory, vestibular, and gustatory systems
___ 6. Special afferent	

i. to the central nervous system
ii. to the muscles of the jaw, facial expression, pharynx, and larynx
iii. to the heart, smooth muscles, and glands
iv. to most of the skeletal muscles

7. By what name is the efferent (motor) division of the visceral nervous system more commonly known? _____

II. Cranial Nerves (453)

1. How many pairs of cranial nerves does a person have? _____

2. Which are the nerves of the cerebrum? _____

3. Which are the nerves of the brainstem? ___ through ___.

4. From what part of the brain do the motor (efferent) fibers of cranial nerves emerge?

   _____

5. Where are the cell bodies of these motor neurons located? _____

6. What two functions do the axons of these neurons have?

   a. _____

   b. _____

7. Where are the sensory (afferent) neuron cell bodies located? _____

8. With what do the axons of the sensory cranial neurons synapse? _____

9. With what sorts of functions are the cranial nerves associated? _____

10. With what sense is cranial nerve I associated? _____

11. Approximately how many neurons fare in cranial nerve I ? _____. Where are
    their cell bodies located? _____

____ 12. The neurons of cranial nerve I are   a. myelinated multipolar
                                                 b. unmyelinated unipolar
                                                 c. myelinated bipolar
                                                 d. myelinated unipolar
                                                 e. unmyelinated bipolar

13. The axons of olfactory neurons synapse with axons of other neurons in the _____ .

14. What is unique about olfactory neurons? _____

15. With what sense is cranial nerve II associated? _____

16. Where do the axons of the medial half of the retinas cross over to the opposite side of the
    brain? _____

17. Where do most axons of the optic tract terminate? _____

18. Why is the retina considered to be a part of the brain? _____

19. What is unusual about the retina, as compared with other parts of the brain? _____

20. Why is the optic nerve technically a tract rather than a nerve? _____

21. Which cranial nerves are the extraocular motor nerves? _____

22. Decribe the function of the oculomotor nerve. _____

23. Which of these functions do the parasympathetic fibers serve? _____

24. Where is the origin of the oculomotor nerve located? _____

25. What muscle does the trochlear nerve innervate? _____

26. Complete the following table, pertaining to the trigeminal nerve.

Name of Branch	Passage	Function
	Superior orbital fissure	
Mandibular nerve		
	Foramen rotundum	

27. What function does cranial nerve VI serve?_____

28. Where are the cell bodies of the facial nerve located?

    a. sensory _____

    b. motor _____

29. On the figure below, identify the following: motor nucleus of nerve VII, sublingual gland, submandibular gland, submandibular ganglion, pterygopalatine ganglion, superior salivatory nucleus, lacrymal gland.

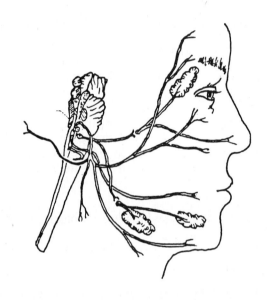

30. What are the two components of cranial nerve VIII?

    a. _____

    b. _____

31. What information does the cochlear nerve convey? _____

32. Where do the axons of the cochlear nerve terminate? _____

33. What information does the vestibular nerve convey? _____

34. Where do its axons terminate? _____

35. Complete the following table.

Nerve/Type	Function	Origin	Distribution
		Medulla oblongata, cervical spinal cord	
			Tongue muscles
	Taste, other sensations of tongue, secretion of saliva, swallowing		
Vagus/mixed			

Match the following.

___ 36. Vestibulocochlear	a. I
___ 37. Olfactory	b. II
___ 38. Trigeminal	c. III
___ 39. Vagus	d. IV
___ 40. Oculomotor	e. V
___ 41. Hypoglossal	f. VI
___ 42. Abducens	g. VII
___ 43. Optic	h. VIII
___ 44. Accessory	i. IX
___ 45. Trochlear	j. X
___ 46. Glossopharyngeal	k. XI
___ 47. Facial	l. XII

48. What are the sensory neurons derived from? _____

49. What neurons does the otic vesicle give rise to? _____

50. What vesicles give rise to the olfactory and optic nerves?

    a. _____

    b. _____

51. Identify which nerves the following give rise to.

    a. neuroblasts of the lateral plate of the brainstem _____

    b. neuroblasts of the basal plate of the neural tube _____

    c. neuroblasts in the basal plate of the cervical spinal neural tube _____

    d. neuroblasts of the cranial neural crests _____

III. Structure and Distribution of Spinal Nerves (463)

1. What are the two roots of the spinal nerves?

    a. _____

    b. _____

2. What is formed by the meeting of the two roots? _____

3. Through what space do most spinal nerves pass? _____

4. For each of the following nerves, identify the two vertebrae proximal to its exit (e.g., C3: exits between C2 and C3).

    a. C4 _____

    b. C7 _____

    c. C8 _____

    d. T1 _____

    e. L1 _____

    f. L2 _____

5. From the outside in, what are the three layers of connective tissue that encase nerve fibers?

   a. _____

   b. _____

   c. _____

6. What is a bundle of axons called? _____

Match the following pairs.

___ 7. Innervates vertebrae, spinal meninges, and spinal blood vessels

___ 8. Innervates skin of back, back of head, and the tissues and intrinsic muscles of back

___ 9. General visceral afferent system and general visceral efferent system

___ 10. Innervates skin, muscles, and tissue of the neck, chest, abdominal wall, limbs, and pelvic area

a. dorsal ramus
b. ventral ramus
c. meningeal ramus
d. rami communicantes

11. Which spinal nerves do not arrange themselves into plexuses? _____

12. Which part of the spinal nerves form the plexuses? _____

13. What are the names and functions of the three branches of nerves coming from the cervical plexus?

    a. _____

    b. _____

    c. _____

14. What areas of the body do the nerves of the brachial plexuses innervate? _____

15. The ventral rami of what nerves form the lumbar plexus? _____

16. What areas of the body does the lumbar plexus innervate? _____

17. The ventral rami of what nerves form the sacral plexus? _____

18. Give the name and innervation sites of the three major nerves that are derived from the sacral plexus.

    a. _____

    b. _____

    c. _____

19. What nerves form the coccygeal plexus? _____

20. What nerves constitute the intercostal nerves? _____

21. What do the various intercostals innervate? _____

22. What is a dermatome, and what is its clinical significance? _____

23. From what embryonic structures do the neurons of the spinal nerves originate? _____

    _____

Match the following neurons with the neuroblasts from which they form.

___ 24. Preganglionic sympathetic neurons
___ 25. Sensory neurons
___ 26. Sacral preganglionic parasympathetic neurons
___ 27. Motor neurons that innervate skeletal muscle

a. neuroblasts of the neural crest
b. neuroblasts that migrate from the matrix
layer to the basal plate of the neural tube
c. neuroblasts that migrate from the matrix
layer to the lateral plate of the neural tube

V. Degeneration and Regeneration of Peripheral Nerve Fibers (470)

1. What happens to a regenerating sprout that fails to reach a nerve ending compatible with
the sprout's functional role? _____

2. What is the function of laminin and NCAMs in the regeneration of nerve cells? _____

_____

3. How long does the process of nerve regeneration take? _____

VI. When Things Go Wrong (472)

1. What results when the facial nerve swells within the facial canal? _____

2. List seven conditions or diseases with which peripheral neuritis is associated.

a. _____

b. _____

c. _____

d. _____

e. _____

f. _____

g. _____

Key
Terms

cranial nerves 453
degeneration 470
dermatome 469
endoneurium 463
epineurium 463
fascicle 463

intercostal nerves 469
mixed nerve 463
motor (efferent) neurons 451
motor nuclei 453
perineurium 463
plexus 465

ramus 465
regeneration 471
sensory (afferent) neurons 451
sensory ganglia 453
spinal nerves 463

Post Test

Matching

____ 1. Neurons that convey impulses directly to the brain or spinal cord (CNS)

____ 2. Neurons that convey impulses away from the CNS

____ 3. Nerves that include both motor and sensory components

____ 4. Axons that innervate skeletal muscle cells

____ 5. Nerves that innervate a leg

____ 6. Neurons that convey action potentials from the heart

____ 7. Neurons that convey action potentials to the stomach

____ 8. Neurons that carry impulses from the olfactory epithelium to the olfactory bulb

____ 9. The autonomic nervous system

____ 10. Neurons that have no cell bodies outside the CNS

a. somatic afferent
b. visceral afferent
c. either somatic or visceral afferent
d. somatic efferent
e. visceral efferent
f. either somatic or visceral efferent
g. mixed

____ 11. I
____ 12. II
____ 13. III
____ 14. IV
____ 15. V
____ 16. VI
____ 17. VII
____ 18. VIII
____ 19. IX
____ 20. X
____ 21. XI
____ 22. XII

a. olfactory
b. hypoglossal
c. trochlear
d. oculomotor
e. vagus
f. abducens
g. facial
h. optic
i. accessory
j. trigeminal
k. vestibulocochlear
l. glossopharyngeal

Using the same lettered list for questions 11–22, match the following.

____ 23. Has a bulbar root

____ 24. Has a cochlear branch

____ 25. Has a mandibular branch

____ 26. Has the geniculate ganglion

____ 27. Has a maxillary branch

____ 28. Has a spinal root

Label the diagram, using the letters for reference.

___ 29. olfactory
___ 30. optic
___ 31. oculomotor
___ 32. trochlear
___ 33. Trigeminal
___ 34. abducens
___ 35. facial
___ 36. vestibulocochlear
___ 37. glossopharyngeal
___ 38. vagus
___ 39. accessory
___ 40. hypoglossal

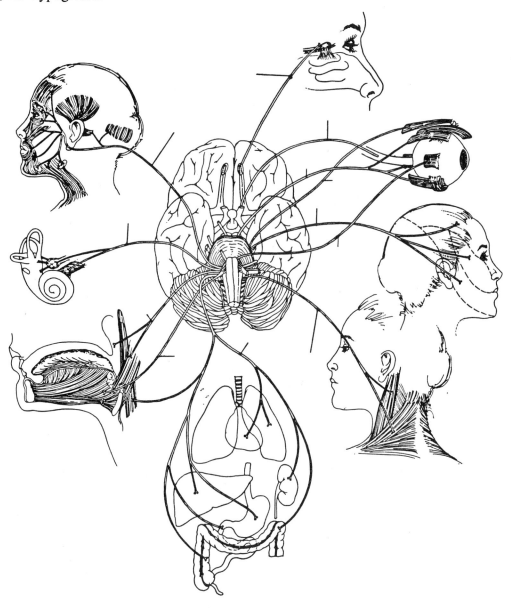

Associate the nerves with the conditions or processes.

____ 41. Vertigo, dizziness
____ 42. Speaking and swallowing
____ 43. Chewing
____ 44. Smiling, scowling
____ 45. Humming, singing

a. mandibular nerve
b. vestibular nerve
c. facial nerve
d. accessory nerve
e. hypoglossal nerve

Identify the nerve plexuses indicated in the diagram.

____ 46. lumbar plexus
____ 47. cervical plexus
____ 48. sacral plexus
____ 49. coccygeal plexus
____ 50. brachial plexus

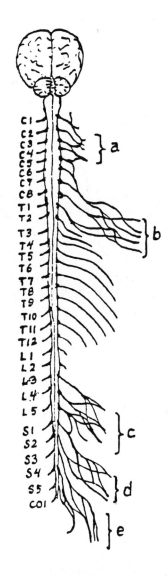

Integrative Thinking

1. Identify the cranial nerve (or component thereof) that is associated with the following functional disorders.

(a) Paresthesis, tic douloureux

(b) Tinnitis, nystagmus, and vertigo

(c) Difficulty speaking, difficulty swallowing

(d) Drooping eyelid, squinting, double vision

2. The "hangman's noose" is characterized by a rather large knot located at the beginning of the loop (see figure). What is the purpose of a knot so big, and what are its immediate effects when the trapdoor under the victim is suddenly sprung?

Your Turn

Imagine that you are about to take a stroll. Carry a small pad of paper and a sharpened pencil with you on your imaginary stroll and wherever it says "Record" in these instructions, mark down the names of the nerves (whether cranial or spinal) and their course into, through, and out of the CNS as they apply.

Start standing at the curb. You look to the left for cars. Record: (a) turning your head, (b) turning your eyes. You look to the right. Record : ditto (a) and (b). You cross the street, looking up at the traffic light. Record: (a) tipping your head upward, (b) lifting your gaze, (c) retaining your balance as you start to trip at an uneven spot in the street. You jump when startled by a horn right behind you. Record your reflex reaction. Now you made it safely across and you start your stroll down the avenue. But wait! What is that tantalizing odor? Record: smells like something you love, and you haven't had your lunch yet. You start to salivate, and you swallow your saliva. Record: (a) salivation, (b) swallowing. Just then you hear a familiar voice call your name. It's Leslie! You respond with a big smile, widening your eyes, and calling, "Leslie! Come join me for lunch!" Record: (a) hearing, (b) smiling, (c) widening your eyes, (d) calling back. You take Leslie's hand and "Ouch!", you get an electric shock. Record: your reflex reaction. You both laugh; it must be the dry weather. You enter the restaurant and have a delightful lunch. Record: Is Leslie a man or a woman?

16 The Autonomic Nervous System

Active Reading

Introduction

1. In what three effectors does the autonomic nervous system (ANS) produce involuntary responses?

   a. _____

   b. _____

   c. _____

2. What are the three divisions of the autonomic nervous system?

   a. _____

   b. _____

   c. _____

3. What is another name for the autonomic nervous system? _____

4. What is the primary function of the ANS? _____

5. Why is the ANS labeled a motor system? _____

6. From what systems does the ANS derive its sensory input? _____

I. Structure of the Peripheral Autonomic Nervous System (479)

Match the following pairs.

____ 1. Activity can be independent of CNS   a. sympathetic

____ 2. Active when the body operates   b. parasympathetic
under stressful conditions   c. enteric

____ 3. Active when there is no bodily
stress

4. Which division of the ANS derives from the thoracolumbar outflow? _____

5. What seven nerve pairs make up the craniosacral outflow of the peripheral ANS?

    a. _____

    b. _____

    c. _____

    d. _____

    e. _____

    f. _____

    g. _____

___ 6. Which of the following is not true of a preganglionic neuron?

    a. cell body is in the brainstem or spinal cord
    b. unmyelinated axon courses through cranial or spinal nerve
    c. axon terminates in synapses in an autonomic ganglion
    d. first neuron in a two-neuron linkage

___ 7. Which of the following is not true of a postganglionic neuron?

    a. cell body is located in an autonomic ganglion
    b. unmyelinated axon courses through nerves and plexuses
    c. axon terminates directly or by way of gap junction
    d. axon terminates on cardiac, smooth, or skeletal muscle, or on a gland

8. Which of the following statements is true (T) and which is false (F)?

    a. Pre- and postganglionic neurons synapse in the CNS. ___

    b. Pre- and postgangiolnic neurons synapse outside the brain and spinal cord. ___

    c. Pll autonomic ganglia are clusters of cell bodies and dendrites. ___

    d. Pll autonomic ganglia are clusters of postganglionic cell bodies and their synapses with preganglionic axons. ___

9. List the three groups of autonomic ganglia, and identify whether each forms part of the sympathetic or parasympathetic system.

    a. _____

    b. _____

    c. _____

10. Describe the location and function of the sympathetic trunk ganglia. _____

_____

11. Name and describe the location of the three major prevertebral ganglia.

    a. _____

    b. _____

    c. _____

12. Where are the terminal ganglia located? _____

13. Complete the following table.

Plexus	Location	Effect
	On the aorta	
enteric plexus		
	Posterior to each lung	
		Regulates the heart
	Behind the stomach	

14. Describe the location and function of the splanchnic nerves. _____

15. Where does the sympathetic division of the ANS arise? _____

16. From which spinal nerves does the thoracolumbar outflow arise? _____

17. What three pathways are available to a sympathetic preganglionic axon after entering the sympathetic trunk?

    a. _____

    b. _____

    c. _____

18. With how many postganglionic neurons does a typical preganglionic sympathetic neuron synapse? _____

19. Label the diagram below as completely as you can.

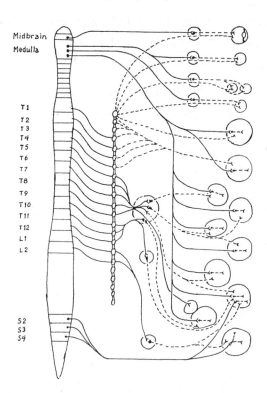

20. What neurotransmitter do preganglionic neuron terminals release? _____

21. What enzyme inactivates it? _____

22. What is the neurotransmitter released by the majority of postganglionic sympathetic neurons? _____

23. What two enzymes inactivate it? For each, identify whether the inactivation takes place within the nerve terminal or in the area of the synaptic cleft.

    a. _____

    b. _____

24. Why is the sympathetic division known as the adrenergic division? _____

25. Why is the parasympathetic division known as the cholinergic division? _____

Match the following catecholamines with their usual effect.

____ 26. Alpha receptors   a. excite only
____ 27. Beta receptors   b. some excite, others inhibit

28. Where do the preganglionic neurons of the parasympathetic division originate? _____

29. In general, are axons of preganglionic parasympathetic neurons longer than the postganglionic ones? _____

30. What four nerves make up the cranial parasympathetic outflow?

    a. _____

    b. _____

    c. _____

    d. _____

31. The roots of which nerves convey the sacral parasympathetic outflow? _____

32. Identify and describe the location of the two types of cholinergic receptors.

    a. _____

    b. _____

33. Comparing the effects of the sympathetic and parasympathetic divisions, which have a shorter duration? _____

34. How do you account for your answer to the preceding question? _____

35. Which nerves are intrinsic to the gut? _____

36. Which divisions of nerves are extrinsic to the gut? _____

37. What effect do the sympathetic and parasympathetic divisions have on the enteric division? _____

38. How many neurons make up the enteric division? _____

39. What functional dependency does the enteric division of the autonomic nervous system have on the central nervous system? _____. How do we know this to be the case? _____

40. Where are the neurons of the enteric division located for the most part? _____

41. What three types of neurons make up the enteric division?

    a. _____

    b. _____

    c. _____

42. List seven neurotransmitters of the enteric division.

    a. _____

    b. _____

    c. _____

    d. _____

    e. _____

    f. _____

    g. _____

43. Which organs you labeled in question 19 do not receive dual innervation by both sympathetic and parasympathetic divisions? _____

    _____

44. Indicate what the effect of sympathetic and of parasympathetic stimulation is on each of the following anatomical stuctures — effects such as increase vs. decrease, dilation vs. constriction, stimulation vs. inhibition.

Structure	Sympathetic	Parasympathetic
Salivary glands		
Peristalsis		
Bronchial tubes		
Gallbladder		
Pupil of eye		
Urinary bladder		
Sweat glands		
Gut sphincters		
Lacrimal glands		

II. Autonomic Control Centers in the Central Nervous System (489)

1. What is the highest and main subcortical neural center of the ANS? _____

2. Why is it considered the coordinating center of the ANS? _____

3. What ANS regulatory centers are present in the brainstem and spinal cord? _____
   _____

4. What areas of the hypothalamus are generally associated with the parasympathetic division of the ANS? _____. With the sympathetic division? _____
   _____

5. What effects does stimulation of the limbic system produce? _____

6. What does a visceral reflex innervate? _____

7. List the five components of a visceral reflex arc.

   a. _____

   b. _____

   c. _____

   d. _____

   e. _____

8. List four examples of reflex arcs in the medulla oblongata.

   a. _____

   b. _____

   c. _____

   d. _____

III. Functions of the Autonomic Nervous System (490)

1. During a high-stress situation, such as a ski race, which division of the ANS is in nearly total command? _____

2. What happens once the stressful event comes to an end? _____

3. What type of activity requires the coordinated, sequential operation of the parasympathetic and sympathetic divisions of the ANS? _____

4. What will happen to the heart if it is deprived of information from the ANS? _____
   Will it continue to function? _____

IV. When Things Go Wrong (491)

1. What symptoms indicate an interruption of either the sympathetic preganglionics or postganglionics to the head? _____

2. Describe the cause and symptoms of achalasia of the esophagus. _____
   _____

3. A lack of which neurons causes Hirschsprung's disease? _____

Key Terms

autonomic ganglia 480
autonomic nervous system (ANS) 478
autonomic plexuses 480
enteric division of ANS 488
neural center 489
parasympathetic division of ANS 487
paravertebral ganglia 480

Peripheral autonomic nervous system (PANS) 479
postganglionic neuraon 479
preganglionic neuron 480
sympathetic division of ANS 481
terminal ganglia 480
thoracolumbar division 479

Post Test

Multiple Choice

____ 1. The autonomic nervous system is a division of the

a. somatic nervous system
b. visceral nervous system
c. central nervous system
d. enteric nervous system
e. none of the above

____ 2. The term thoracolumbar outflow applies to the

a. peripheral autonomic nervous system (PANS)
b. parasympathetic division of the PANS
c. prevertebral ganglionic chain
d. thoracolumbar plexus
e. sympathetic division of the PANS

____ 3. The celiac, superior mesenteric, and inferior mesenteric ganglia are

a. prevertebral ganglia
b. sympathetic ganglia
c. ganglia of the sympathetic trunk
d. cranial autonomic ganglia
e. thalamic basal ganglia

____ 4. Terminal ganglia are

a. diffuse ganglia close to or embedded in certain visceral organs
b. postganglionic fibers of either sympathetic or parasympathetic divisions of the PANS
c. degenerating autonomic ganglia
d. temporary neurons of the PANS
e. synonymous with sympathetic ganglia

____ 5. The preganglionic neuron cell bodies of the sympathetic division are located in the

a. gray rami communicantes
b. white rami communicantes
c. lateral gray horn of the spinal cord
d. anterior gray horn of the spinal cord
e. dorsal (superior) root ganglia

Identify the structures shown in the diagram below, using the letters for reference.

____ 6. autonomic ganglion
____ 7. neurotransmitter molecules
____ 8. preganglionic neuron cell body
____ 9. postganglionic neuron cell body
___ 10. unmyelinated postganglionic fiber

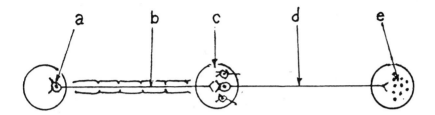

Identify the structures specified in the figure below, using the letters for reference.

___ 11. prevertebral ganglion
___ 12. sympathetic trunk
___ 13. lateral horn
___ 14. terminal ganglion
___ 15. preganglionic axon
___ 16. postganglionic axon
___ 17. spinal ganglion
___ 18. gray ramus
___ 19. white ramus
___ 20. anterior (ventral) root of spinal nerve

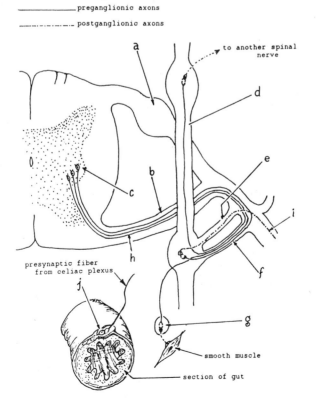

194   Chapter 16

Matching

___ 21. Neurotransmitter for adrenergic receptors

___ 22. Neurotransmitter for most preganglionic neurons of the sympathetic division

___ 23. Deactivated by catechol-o-methyl transferase

___ 24. Major transmitter for postganglionic neurons of parasympathetic division

___ 25. A neurotransmitter of the enteric division of the PANS

___ 26. Neurotransmitter for nicotinic receptors

___ 27. Anticholinergic

___ 28. Deactivated by monoamino oxidase

___ 29. Deactivated by cholinesterase

___ 30. A catecholamine

a. acetylcholine
b. norepinephrine
c. atropine
d. serotonin

___ 31. The main subcortical regulatory center of the ANS

___ 32. Regulatory center for visceral reflex arcs

___ 33. Location of spinal reflex arcs

___ 34. Location of reflex arcs involved in peristalsis

___ 35. Regulates many visceral functions with no central nervous system participation

___ 36. Location of respiratory centers

___ 37. Major location at which reflex centers for homeostasis are coordinated and controlled

___ 38. Location of reflex centers for sensory input from the urinary bladder

___ 39. Nervous system intrinsic to the gut

___ 40. Involved in many behavioral responses, such as those involved in self-preservation, love-making, or child care

a. spinal cord
b. medulla oblongata/brainstem
c. enteric division of the PANS
d. limbic system
e. hypothalamus

Integrative Thinking

1. Visceral reflexes typically occur without one's conscious control or awareness. How do you explain this?

2. The emotions often affect functions that are usually considered to be in the purview of the autonomic nervous system. For example, fear affects heart rate, blood pressure, sweating, intestinal peristalsis, urinary function, respiration, visual acuity, and several other functions. How do you explain this?

Your Turn

Devise tests that the members of a study group can realistically try on each other that (a) will elicit a visceral reflex that (b) can be demonstrated — for example, increased flow of saliva in response to a delectable odor; alteration of pupil size by different intensities of light; change in pulse rate caused by. But, there! You think them up! How is your imagination? Does anybody have access to a stethoscope?

17 The Senses

Active Reading

Introduction

1. With what sensations is the skin involved?

   a. _____

   b. _____

   c. _____

   d. _____

2. What changes can receptors in the circulatory system register?

I. Sensory Reception (496)

1. What are sensory receptors? _____

___ 2. What is sound?

    a. movement of air or other media in a wavelike pattern

    b. vibrations within the ear

    c. a perception of the brain

    d. any or all of the above

3. What is a transducer?

___ 4. Different sensory transducers

    a. convert the same types of stimuli into different kinds of nerve impulses

    b. convert different types of stimuli into different kinds of nerve impulses

    c. convert different types of stimuli into the same kind of nerve impulse

    d. all of the above

5. Contrast receptor potentials and action potentials. _____

_____

6. List three features common to all sensory receptors.

    a. _____

    b. _____

    c. _____

Match the following pairs.

  ___ 7. Located in organs with motor   a. exteroceptors
          innervation from the autonomic   b. teleceptors
          nervous system   c. interoceptors
  ___ 8. Involved in sensing position of body   d. proprioceptors
          in space and relative location of
          different parts
  ___ 9. Detect environmental changes that
          occur some distance from the body
  ___ 10. Respond to external environmental
          stimuli that affect the skin directly

11. Which are the special senses?

    a. _____

    b. _____

    c. _____

    d. _____

12. Why are they special? _____

13. What are some of the general senses?

    a. _____

    b. _____

    c. _____

    d. _____

    e. _____

    f. _____

14. What is kinesthesia? _____

15. Name six classes of receptors based on the kind of stimulus to which they respond.

    a.

    b.

    c.

    d.

    e.

    f.

Match the following pairs.

____ 16. Most widespread of all the sensory receptors   a. thermoreceptors
____ 17. Respond to the visual stimuli of visible light waves   b. nociceptors
____ 18. Respond to potentially harmful stimuli that produce pain   c. chemoreceptors
____ 19. Respond to chemical stimuli that result in taste and smell   d. photoreceptors
____ 20. Mechanoreceptors that respond to changes in blood pressure   e. mechanoreceptors
____ 21. Respond to temperature changes.   f. baroreceptors

22. Contrast free nerve endings and encapsulated endings. _____

23. List six important encapsulated receptors.

     a. _____

     b. _____

     c. _____

     d. _____

     e. _____

     f. _____

24. What distinguishes a nerve impulse for sight from one for sound? _____

_____

II. General Senses (498)

1. Complete the following table.

Sense	Stimulus(i)	Receptor(s)
Light touch		
	Deformation of skin	
	External temperature above or below that of skin	
Pain		
	Continuing periodic change in displacement	
Itch and tickle		
	Position of body parts relative to each other	

2. What is stereognosis? _____

Match the following neural pathways with their respective senses.

____ 3. Light touch
____ 4. Touch-pressure
____ 5. Temperature
____ 6. Pain
____ 7. Conscious proprioception

a. lateral spinothalamic tract
b. anterior spinothalamic tract
c. indirect spinoreticulo-thalamic pathway
d. dorsal column-medial lemniscal pathway

8. In what area of the brain are unconscious proprioceptive impulses received? _____

9. What sensations does the trigeminothalamic tract convey? _____

III. Taste (Gustation) (504)

1. What are the three main types of lingual papillae?

   a. _____

   b. _____

   c. _____

2. What are gustducin and cyclic guanosine monophosphate (cGMP)? _____

3. Describe the interaction of different taste receptors. _____

4. What three nerves carry impulses of taste sensations?

   a. _____

   b. _____

   c. _____

IV. Smell (Olfaction) (505)

1. How many different odors can a typical adult detect? _____

2. Do all chemicals have an associated smell? _____

3. In the figure below, identify the following: olfactory epithelium, receptor cell, supporting cell, basal cell, olfactory vesicle, cilia, olfactory gland, olfactory bulb, olfactory tract.

(enlarged)

4. Where are odors first processed? _____

5. Where is the primary olfactory cortex? _____

6. What is unique about the projection of olfactory fibers as compared to those of other senses? _____

7. Identify the role of cAMP in the olfactory process. _____

8. What is functionally unusual about olfactory receptor cells? _____

9. Into what do cells of the nasal placodes differentiate? _____

V. Hearing and Equilibrium (507)

1. What nerve innervates the auditory apparatus? _____

2. What nerve innervates the vestibular apparatus? _____

3. What are the two nerves collectively known as? _____

4. What are the auditory ossicles? _____

5. How are the ossicles held in place and attached to each other? _____

6. In the figure below, identify the following: auricle, external auditory canal, tympanic membrane, tympanic cavity, auditory (Eustachian) tube, malleus, incus, stapes, semicircular canals, cochlea, spiral organ, vestibular nerve, cochlear nerve.

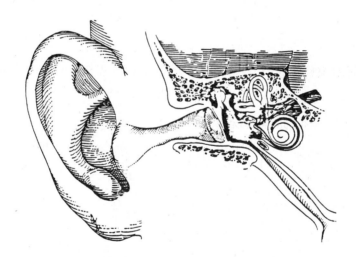

7. Trace the avenue by which the middle ear acquires an infection that results in otitis media. _____

_____

8. By what mechanism is the intensity of sound modified? _____

9. Where in the ear are the sound receptor cells? _____

10. What are perilymph and endolymph, and where are they located? _____

_____

11. In the figure below, identify the superior, posterior, and lateral semicircular canals, the utricle, saccule, scala vestibuli, and scala tympani.

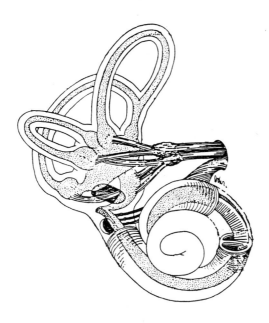

12. Briefly describe the function of the spiral ganglion. _____

13. Define the terms frequency, timbre, intensity, and pitch, and comment on how each is determined. _____

_____

14. Why does a lesion of the auditory pathway on one side result in a decrease of hearing acuity on both, but predominantly the opposite, side of the body? _____

15. What are the three main components of the vestibular apparatus?

    a. _____

    b. _____

    c. _____

16. What excites a hair cell? _____

17. Describe how the relative densities of otoconia and endolymph result in the bending of hairs of a hair cell. _____

18. Why is it it important that the three semicircular ducts lie on different axes? _____

_____

19. What three important roles do inputs from the vestibular receptors serve?

    a. _____

    b. _____

    c. _____

VI. Vision (522)

1. In the figure below, identify the following: sclera, cornea, choroid, ciliary body, lens, pupil, retina, optic nerve, macula lutea, fovea centralis, peripheral retina, posterior chamber, anterior chamber, scleral venous sinus, vitreous chamber.

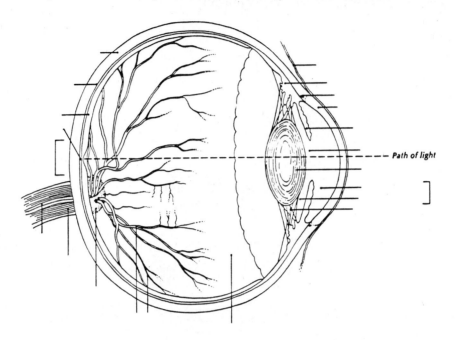

Path of light

2. Describe the location, composition, and function of aqueous humor. _____

    _____

3. Describe the location, composition, and function of vitreous humor. _____

    _____

4. Name and comment on the function of the various kinds of cells that make up the neuroretina. _____

5. What are the functions of

    a. eyelashes _____

    b. eyebrows _____

6. What are the two parts of the conjunctiva?

    a. _____

    b. _____

7. What are the four functions of the secretions of the lacrimal gland?

    a. _____

    b. _____

    c. _____

    d. _____

8. Identify the following muscles as either extrinsic or intrinsic.

    a. rectus muscles _____

    b. ciliary muscle _____

    c. superior oblique muscle _____

    d. inferior oblique muscle _____

    e. circular muscle _____

    f. radial muscle _____

9. Which of the muscles in Question 8 are smooth? _____

___ 10. Light waves

    a. travel straight through a medium of consistent density
    b. bend when passing between media of different densities
    c. after passing through a lens, converge on a focal point
    d. all of the above

___ 11. Accommodation

    a. adjusts the position of the focal point by changing the shape of the cornea
    b. adjusts the position of the focal point by changing the shape of the lens
    c. regulates the amount of light entering the eye by adjusting the size of the pupil
    d. all of the above

___ 12. Convergence

    a. marks the successful operation of the lens to make light waves converge on a single focal point
    b. marks the slight inward turn of the eyeballs so that their lines of sight converge on a single object
    c. marks the reduction in size of the pupil so that light waves must converge through a narrow aperature
    d. all of the above

Match the following pairs.

___ 13. Perceive shades of color   a. rods
___ 14. Perceive black and white   b. cones

15. Describe the role of the photopigment rhodopsin in the conversion of electromagnetic energy into the hyperpolarization and eventual action potential of rod cells. _____

_____

16. What role does vitamin A play in vision? _____

17. Describe how cones differ from rods, resulting in the ability to discriminate a wide range of colors. _____

18. Explain the physiology of the two stages of dark adaptation.

    a. _____

    b. _____

19. What are the four half-fields of vision?

    a. _____

    b. _____

    c. _____

    d. _____

20. Where do the two optic nerves meet? _____

21. Where do the optic tracts terminate? _____

22. What is unique about the following, as compared to other parts of the brain and/or nervous system?

    a. retina _____

    b. optic nerve _____

23. What are the five main classes of neurons in the retina? Identify whether each forms a direct pathway to the brain.

    a. _____

    b. _____

    c. _____

    d. _____

    e. _____

24. Identify the classes of cells illustrated in the diagram below, using letters for reference.

    a. _____

    b. _____

    c. _____

    d. _____

    e. _____

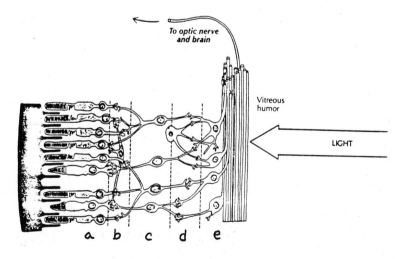

Match the following pairs.

___ 25. On center, off surround
___ 26. Off center, on surround

   a. stimulation of center inhibits ganglion firing
   b. stimulation of center increases rate of ganglion firing

27. Describe the roles of the following structures in the development of the eye.

   a. optic vesicles _____

   b. optic stalk _____

   c. optic cup _____

   d. lens vesicle_____

   e. optic fissures _____

VII. The Effects of Aging on the Senses (539)

1. What causes impairment of eyesight as one ages? _____

2. Define or describe the following conditions.

   a. presbyopia _____

   b. cataract _____

   c. glaucoma _____

3. What is the leading cause of blindness in people over 65 years of age? _____

4. How are hearing, smell, and taste affected by aging? _____

VIII. When Things Go Wrong (539)

1. Give the cause(s) and symptom(s) for the following.

   a. otosclerosis _____

   b. labyrinthitis _____

   c. Ménière's disease _____

   d. motion sickness _____

   e. otitis media _____

2. What procedures can help people with a detached retina? _____

3. What disease causes portions of the lens to become opaque? _____

4. What is "pinkeye," and how is it caused? _____

5. Describe the chain of events in glaucoma that lead to blindness. _____

_____

6. Describe the sensations associated with nystagmus. _____

_____

Key Terms

accomodation 528, 529
adaptation 532
auditory apparatus 509
auditory ossicles 510
auditory tube 510
baroreceptor 498
bipolar cell 535
bulbous corpuscles 500
chemoreceptor 497
choroid 521
ciliary body 521
cochlea 512
cone 521
convergence 528, 531
cornea 520
corpuscles of Ruffini 500
dynamic equilibrium 516
encapsulated ending 498
external auditory canal 509
free nerve ending 498, 500
general senses 497, 498

gustation 504
hair cells 516
horizontal cell 535
iris 521
labyrinth 512
lacrimal apparatus 527
lamellated corpuscles 500
lens 521
light touch 497, 498
mechanoreceptor 497
nociceptor 497
olfaction 505
olfactory bulbs 507
olfactory receptor cells 506
optic chiasma 535
photoreceptor 497, 535
proprioception 502
pupil 521
receptor 496
receptor potential 496, 502
retina 521

retinal layer 521
rhodopsin 531
rod 521
saccule 512
sclera 520
semicircular canals 512
semicircular ducts 512
special senses 497
spiral organ (of Corti) 512
static equilibrium 516
stereognosis 502
tactile (Meissner's) corpuscles 500
tactile (Merkel's) corpuscles 500
taste buds 504
thermoreceptor 497
touch-pressure 497
tympanic cavity 510
tympanic membrane 510
utricle 512
vestibular apparatus 509, 516
vestibule 512

Post Test

Multiple Choice

_____ 1. An object or device that converts one form of energy into another is known as a

a. transducer
b. transformer
c. transmitter
d. transfixer
e. translater

_____ 2. Sensory receptors are organs or organ components that directly convert specific kinds of stimuli into

a. corresponding kinds of nerve impulses
b. specific kinds of neural transmitters
c. nonspecific kinds of neural transmitters
d. nerve impulses that are all alike
e. efferent impulses

_____ 3. Receptors for pain and temperature are

   a. encapsulated nerve endings
   b. free nerve endings
   c. lamellated corpuscles
   d. baroreceptors
   e. proprioceptors

4. Arrange the following items in the order in which they normally occur: (__) > (__) > (__) > (__) > (__)

   a. Na$^+$ flows in, K$^+$ flows out
   b. threshold stimulus
   c. action potential in afferent fiber
   d. receptor potential
   e. generator potential

_____ 5. In general, sensory nerves that convey highly localized and discriminative sensations are, as compared to other afferents,

   a. larger
   b. more heavily myelinated
   c. faster conducting
   d. no different
   e. a, b, and c are all applicable

Matching

_____ 6. Corpuscles of Ruffini
_____ 7. Iris
_____ 8. Temperature receptors
_____ 9. Krause's corpuscles
_____ 10. Blood pressure receptors
_____ 11. Gustoreceptors
_____ 12. Pacinian corpuscles
_____ 13. Olfaction
_____ 14. Nociceptors
_____ 15. Posture and balance

   a. thermoreceptors
   b. chemoreceptors
   c. mechanoreceptors
   d. baroreceptors
   e. photoreceptors
   f. more than one of these

_____ 16. The basic embryological structure from which the olfactory organ differentiates
_____ 17. Parent cells to the olfactory epithelium; sprouts axons that penetrate the telencephalon
_____ 18. Cells whose axons lengthen and form sensory tracts
_____ 19. Cells of this structure synapse with cells of the cerebral hemispheres, forming a pair of prominences
_____ 20. Bundles of axons formed by the secondary neurosensory cells
_____ 21. The final connecting structure between the olfacory epithelium and the brain proper

   a. primary neurosensory cells
   b. secondary neurosensory cells
   c. olfactory bulb
   d. nasal (olfactory) placode
   e. olfactory tract
   f. olfactory nerve

Identify the structures illustrated in the figure below.

___ 22. Scala tympani
___ 23. Tectorial membrane
___ 24. Outer hair cells
___ 25. Spiral ganglion
___ 26. Cochlear duct
___ 27. Scala vestibuli
___ 28. Basilar membrane
___ 29. Vestibular membrane

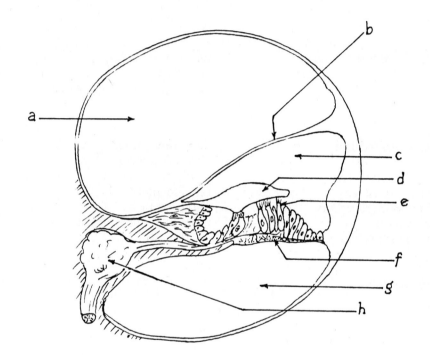

Matching

___ 30. A body appendage composed of a thin plate of elastic cartilage covered by a close-fitting layer of skin.
___ 31. The eardrum.
___ 32. Crystals of calcium carbonate in the macula.
___ 33. A canal connecting the middle ear chamber with the nasopharynx.
___ 34. The outer ear canal.
___ 35. Earwax.
___ 36. A bony ring that supports the eardrum.

a. auditory tube
b. auricle
c. cerumen
d. tympanic annulus
e. external auditory meatus
f. tympanic membrane
g. otoconia

Multiple Choice

___ 37. The range of sound intensity that is "comfortable" to live with on a regular basis, and is not likely to cause any damage to hearing (in decibels)

    a. 75–110
    b. 90–120
    c. 100–150
    d. 5–40
    e. 0–70

___ 38. The range of sound frequencies audible to most human adults is around (in cycles per second)

    a. 15–40,000
    b. 20–35,000
    c. 30–20,000
    d. 40–10,000
    e. 45–900

___ 39. A common cause of hearing loss is

    a. calcification of the tympanic membrane
    b. increased tautness of the basilar membrane
    c. degeneration of the stereocilia (hair cells)
    d. gelation of the tectoral membrane
    e. reduced cerumen secretion

___ 40. The mechanism of the static equilibrium response depends on the difference in density between the

    a. otoconia and the endolymph
    b. the endolymph and the perilymph
    c. the otoconia and the perilymph
    d. the stereocilia and kenocilia
    e. the ampulla and the cupula

Matching

___ 41. Ciliary body
___ 42. Aqueous humor
___ 43. Fovea
___ 44. Sclera
___ 45. Iris
___ 46. Retina
___ 47. Choroid
___ 48. Lens
___ 49. Cornea
___ 50. Macula lutea

    a. supporting layer
    b. vascular layer
    c. photosensitive layer
    d. chambered component

Matching

___ 51. A reddish, photosensitive pigment
___ 52. The neurotransmitter to which retina neurons respond
___ 53. The protein that activates phosphodiesterase
___ 54. The smallest unit of light energy
___ 55. A light-absorbing molecule that is coupled with scotopsin
___ 56. In the dark, it binds to sodium channels in the surface membrane of rod cells, holding the channels open

    a. photon
    b. glutamate
    c. cGMP
    d. rhodopsin
    e. retinal
    f. transducin

Multiple Choice

____ 57. Light coming from the right side of a person's visual field falls on the

    a. temporal side of the right eye and nasal side of the left eye
    b. temporal side of the left eye and nasal side of the right eye
    c. temporal side of both eyes
    d. nasal side of both eyes
    e. fovea centralis of the right eye

____ 58. At the optic chiasma

    a. the two optic nerves cross and become, unaltered, the left and right optic tracts, respectively
    b. afferent fibers from the temporal sides of both eyes cross and enter one optic tract, while those from the nasal sides enter the other tract
    c. fibers from both horizontal halves and both vertical halves become uniformly mixed so that all are represented equally in both optic tracts
    d. fibers from the temporal sides of both eyes cross over
    e. Fibers from the nasal sides of both eyes cross over

____ 59. The nerve fibers that make up the optic nerve are the axons of

    a. ganglion cells
    b. amacrine cells
    c. bipolar cells
    d. horizontal cells
    e. the rods and cones

____ 60. Left and right optic tracts enter the brain, where they synapse with interneurons in the

    a. cerebral cortex
    b. substantia nigra
    c. cerebellum
    d. lateral geniculate bodies
    e. corpus striatum

____ 61. From there, the interneurons send axons to the primary visual cortex, which is in the

    a. temporal lobe
    b. occipital lobe
    c. parietal lobe
    d. frontal lobe
    e. diencephalon

____ 62. Some afferent fibers from the retina, however, go directly to reflex centers in the

    a. cerebellum
    b. epithalamus
    c. medulla oblongata
    d. lenticular nucleus
    e. superior colliculi in the midbrain

____ 63. The eyes and optic nerves develop from

    a. paired placodes along side the neural tube
    b. paired outpocketings of the forebrain
    c. head mesenchyme
    d. a and b
    e. b and c

___ 64. The choroid and sclera are continuous with the

    a. integument of the head
    b. neurosensory epithelium
    c. arachnoid and dura mater of the brain
    d. myelin sheaths of the optic nerves
    e. none of the above

Integrative Thought

1. Try this in a dimly lit or darkened room: Press your index finger against the nasal side of your eyeball with a gentle massaging motion, so that you see a bright spot or small streak. Where does it appear to be? How do you explain this? Now try it against the lateral (temporal) side of the eyeball. Again, where do the little stars appear to be? What is your hypothesis?

2. The most common clinical test for incipient glaucoma involves measuring momentary flattening of the cornea when it is pressed by a transparent rod or a puff of air. Explain what is being measured, and what it implies.

3. You are an ophthalmologist. An elderly patient comes to you, complaining of difficulty with reading. She says the letters are blurred and indistinct, although the rest of her visual field seems normal. You have her look at the eye chart on the wall. You notice that she is hesitant, and then she turns her head a little and succeeds in reading the larger letters. You tell her not to turn her head, but to look straight at the chart. She cannot read it. What is your immediate diagnosis? Explain. What is the prognosis?

4. Now you are an audiologist. A patient comes to you complaining about the hearing aid he inherited from his recently deceased aunt. "My brother hears with it just fine, but I don't hear a darned thing with it, even when I turn up the volume full," he says. You hold a tuning fork beside his ear. He hears nothing. You hold the shaft of the fork against his skull. He hears it clearly. What is the likely cause of his hearing loss? Can anything be done about it?

Your Turn

1. In anticipation of your group meeting, prepare three solutions and a few Q-tips (two per person): (1) a sugar solution, (2) a salt solution, and (3) something sour, such as vinegar or lemon juice. Moisten a Q-tip with one of these substances and touch it to different areas on the tongue and note where it is tasted. Repeat with the other two. Why did we omit something bitter? (Answer: most bitter substances are poisonous, others are semipoisonous but we take them judiciously as medicines, such as aspirin, steroids, etc. They also tend to initiate the vomiting reflex. Can you think of any evolutionary or adaptive implications here?)

2. Procure a pair of dividers, such as marine navigators and draftsmen use. Or devise a substitute with a couple of sharpened pencils or a pair of steady hands. Starting with a team member's back, press the two points of the divider against the subject's back and ask whether he/she felt one point or two points. Vary the distance between the points, and sometimes use only one point. Repeat a few times, keeping score of the results. With the subject's eyes closed, repeat on the forearm, on the fingertips, perhaps even on the tongue. What do your results tell you about touch discrimination? What do they imply about the relative distribution of touch receptors in different areas of the body?

3. Somebody bring an onion, an apple, and a paring knife to the study group meeting. With your nasal passages closed off in this way, chew on a piece of the onion and then apple, or the other way around. How do they compare? Where does one really "taste" onions? Explain.

18 The Endocrine System

Active Reading

Introduction

1. What are the three functions of the endocrine system?

   a. _____

   b. _____

   c. _____

2. Where do endocrine glands secrete hormones? _____

3. Where do exocrine glands secrete hormones? _____

I. Hormones and Their Feedback Systems (548)

1. What is a hormone? _____

2. Where are hormones effective? _____

3. What is the basis of the distinction between endocrines, paracrines, and autocrines?

   _____

Match the following pairs.

___ 4. Lipid-soluble steroids   a. endoderm-derived glands
___ 5. Derivatives of amino acids   b. ectoderm-derived glands
___ 6. Water-soluble proteins and peptides   c. mesoderm-derived glands

7. What is the fourth kind of hormone? _____

8. Some hormones initiate biochemical reactions within the cells. True or false? ___

9. Regulation of homeostasis occurs primarily through the use of positive feedback systems. True or false? ___

10. Negative feedback systems can involve more than one gland and more than one hormone. True or false? ____

II. Mechanisms of Hormone Control (552)

1. Distinguish first messengers from second messengers. _____

_____

2. Describe the six stages of the fixed-membrane-receptor mechanism.

a. _____

b. _____

c. _____

d. _____

e. _____

f. _____

3. Why is cAMP called "cyclic"? _____

4. What are five mobile-receptor hormones?

a. _____

b. _____

c. _____

d. _____

e. _____

5. Describe the six stages of the mechanism of steroid hormone action.

a. _____

b. _____

c. _____

d. _____

e. _____

f. _____

III. Pituitary Gland (Hypophysis) (553)

1. Where is the pituitary gland located? _____

Match the following.

____ 2. Larger lobe   a. adenohypophysis
____ 3. Posterior lobe   b. neurohypophysis
____ 4. Abundance of functional secretory
          cells
____ 5. Greater supply of large nerve
          endings

6. On the figure below identify the following: hypothalamus, superior hypophyseal artery, infundibular stalk, hypothalamic-hypophyseal portal vessels, sinusoids, adenohypophysis, neurohypophysis, inferior hypophyseal artery.

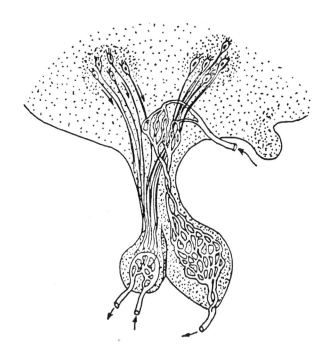

7. What is the real master gland? _____

8. What special kind of cells are the hypothalamic nerve cells? _____

9. What two hormones do the hypothalamic nerve endings secrete? What are their functions?

   a. _____

   b. _____

10. Why is ADH also called vasopressin? _____

11. Complete the following table.

Hormone	Target Organ	Effect(s)
Growth hormone		
Prolactin		
		Follicle cells grow; ovaries secrete estrogen; testes produce sperm
	Thyroid gland	
Melanocyte stimulating hormone		
	Adrenal cortex	
Luteinizing hormone		

IV. Thyroid Gland (558)

1. On the figure below, identify the following: left lobe of thyroid, right lobe of thyroid, isthmus, pyramidal process, thyroid cartilage, hyoid bone, trachea, follicular cells, parafollicular cells, colloid.

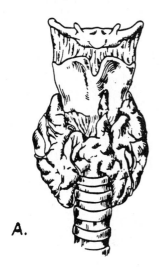

A.

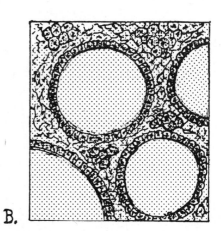

B.

2. What do the two types of thyroid cells secrete?

   a. _____

   b. _____

3. Complete the following table.

Hormone	Site of Production	Target Organ(s)	Effect(s)
Thyroxin			
	Parafollicular cells		
Triiodothyronine			

V. Parathyroid Glands (561)

1. How many parathyroid glands are there? _____

2. Where are they located? _____

3. What are the two secretions of the parathyroid glands?

    a. _____

    b. _____

4. By what three ways does parathormone increase the calcium level in the blood?

    a. _____

    b. _____

    c. _____

5. How does parathormone decrease the concentration of phosphorus in the blood? _____

VI. Adrenal Glands (562)

1. On the figures below, identify the following: kidneys, adrenal glands, adrenal cortex, adrenal medulla, capsule, zona glomerulosa, zona fasciculata, zona reticularis.

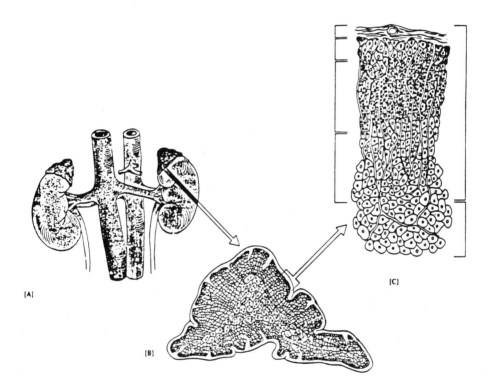

2. Complete the following table.

Hormone	Class	Target Organ(s)	Effect(s)
Cortisol			
	Mineralocorticoid		
Corticosteroid hormone			
Androgen			
Estrogen			

3. Describe the renin-angiotensin-aldosterone mechanism. _____

_____

4. What are chromaffin cells? _____

5. From what tissue are chromaffin cells derived? _____

6. Complete the following table.

Hormone	Effects on Muscles	Effects on Heart	Effects on Blood	Effects on Metabolism
Epinephrine				
Norepinephrine				

7. By what two ways do increased cortisol levels help the body deal with stress?

a. _____

b. _____

8. How does epinephrine help the body deal with stress? _____

VII. Pancreas (568)

1. Why is the pancreas considered a mixed gland? _____

2. How many pancreatic islets does an adult pancreas contain? _____

3. List the four types of islets, together with the hormone each one produces:

a. _____

b. _____

c. _____

d. _____

4. Why is glucagon considered to be a hypoglycemic factor? _____

5. What are the five most important functions of insulin?

    a. _____

    b. _____

    c. _____

    d. _____

    e. _____

6. What do excessive amounts of insulin produce? _____

VIII. Gonads (570)

1. What four hormones are concerned with the development and regulation of male sex organs and male functions?

    a. _____

    b. _____

    c. _____

    d. _____

2. What are the three classes of female hormones, together with their functions?

    a. _____

    b. _____

    c. _____

3. To what extent are the male hormones unique to men, and the female hormones unique to women? _____

IX. Other Sources of Hormones (571)

1. In addition to erythropoietin, 1,25-dihydroxyvitamin D_3, prekallikreins, and prostaglandins, what other hormone do the kidneys produce? _____

2. Where is the pineal gland located? _____

3. What hormone does the pineal gland produce during the night? _____

4. Where is the thymus located? _____

5. What is the main function of the thymus? _____

6. What are some of the hormones and factors of the thymus? _____

7. What hormone do cardiac muscle cells in the two atria of the heart secrete? _____

8. What is the function of this hormone? _____

9. For the following hormones of the digestive system, list the site of production, the target organ, and the effect.

    a. gastrin _____

    b. secretin _____

    c. cholecystokinin _____

10. What two other hormones influence the intestines?

    a. _____

    b. _____

11. What hormones does the placenta produce?

    a. _____

    b. _____

    c. _____

12. What are the eicosanoids? _____

13. Where are prostaglandins found? _____

14. Where are prostaglandins produced? _____

15. Why are prostaglandins not considered true hormones? _____

X.   The Effects of Aging on the Endocrine System (573)

1. Explain the causes of diabetes mellitus. _____

    _____

2. Explain the causes of decreased sexual function with age. _____

    _____

3. List four other effects of an aging endocrine system.

    a. _____

    b. _____

    c. _____

    d. _____

XI. Developmental Anatomy of the Pituitary and Thyroid Glands (574)

1. From what embryonic structure does the neurohypophyseal bud form, and what does the bud ultimately become? _____

2. From what embryonic structure does the anterior pituitary (adenohypophysis) develop?

    _____

3. Where and from what embryonic structure does the thyroid gland originate? _____

    _____

XII. When Things Go Wrong (000)

1. Describe the roles of the pituitary and thyroid glands in the following conditions.

    a. gigantism _____

    b. acromegaly _____

    c. pituitary dwarfism _____

2. What condition results from an undersecretion of ADH? _____

3. List the symptoms and treatment for the following abnormalities.

    a. hypothyroidism _____

    b. hyperthyroidism _____

    c. hypoparathyroidism _____

    d. hyperparathyroidism _____

Match the conditions with the symptoms.

___ 4. Overactive adrenocortical tumor	a. anemia, fatigue, elevated blood potassium, decreased blood sodium
___ 5. Hypoadrenalism	b. redistribution of fat to the face, chest, and abdomen
___ 6. Overproduction of glucocorticoids	c. masculinization in a female, accelerated sexual development in a male

7. Identify the causes, symptoms, and time of onset of the following two conditions.

    a. Type I diabetes _____

    b. Type II diabetes _____

8. Describe the treatment for severe hypoglycemia. _____

_____

9. Define the following terms.

    a. acromegaly _____

    b. cretinism _____

    c. polydipsia _____

    d. goiter _____

    e. myxedema _____

    f. Cushing's disease _____

    g. Addison's disease _____

    h. Graves' disease _____

    i. polyuria _____

    j. polyphagia _____

Key Terms

adenohypophysis 553
adrenal cortex 562
adrenal glands 562
adrenal medulla 566
adrenocorticotropic hormone 558
aldosterone 565
anabolic steroid 564
antidiuretic hormone (ADH) 555
atrial natriuretic peptide 572
calcitonin 560
cholecystokinin 572
cortisol 563
diabetes mellitus 570, 577
endocrine gland 548
epinephrine 566
estrogens 571
fixed-membrane-
receptor mechanism 552
follicle-stimulating hormone 558,
570, 571

gastrin 572
glucagon 568
glucocorticoid 563
gonadocorticoid 566
growth hormone (GH) 557
hormone 548
hypoglycemia 570, 578
hypophysis 553
hypothalamic-hypophyseal
portal system 554
hypothalamus 554
insulin 568
luteinizing hormone 558, 570
mineralocorticoid 565
mobile-receptor mechanism 553
neurohypophysis 553
norepinephrine 566
oxytocin 555

pancreatic islet 568
parathormone 561
parathyroid glands 561
pineal gland 571
pituitary gland 553
progestins 571
prolactin 557
prostaglandins 573
renin 565
secretin 572
target cell 548
testosterone 570
thymus gland 571
thyroid gland 558
thyroid-stimulating hormone 558
thyroxine 559
triiodothyronine 559

Post Test

Matching

____ 1. Cholecystokinin
____ 2. Growth hormone (GH)
____ 3. Somatotropic hormone (STH)
____ 4. Glucocorticoids
____ 5. Glucagon
____ 6. Mineralocorticoids
____ 7. ADH
____ 8. Calcitonin
____ 9. Epinephrine
___ 10. Oxytocin
___ 11. Secretin
___ 12. Insulin
___ 13. FSH
___ 14. Luteinizing hormone
___ 15. PTH

a. neurohypophysis
b. adenohypophysis
c. thyroid gland
d. parathyroid gland
e. adrenal gland
f. digestive system
g. pancreas

___ 16. Insulin
___ 17. Atriopeptin
___ 18. Glucagon
___ 19. Progesterone
___ 20. Aldosterone
___ 21. Norepinephrine
___ 22. Ovary
___ 23. Androgens

a. alpha cells
b. beta cells
c. corpus luteum
d. heart
e. adrenal medulla
f. adrenal cortex

___ 24. Estrogen
___ 25. Insulin
___ 26. Growth hormone
___ 27. Epinephrin
___ 28. Histamine
___ 29. Prostaglandins
___ 30. Thyrotropin-releasing hormone

a. steroid
b. biogenic amine
c. protein/peptides
d. eicosanoid

Miscellaneous/Multiple Choice

___ 31. Arrange the following molecules in their proper sequence, starting with the corticotropin-releasing hormone (CRH) and ending with the secretion of a glucocorticoid. (_a_)>(___)> (___)>(___)>(___)>(___)>(___) > (___)>(_i_)

a. CRH
b. adrenal cyclase
c. hormone receptor
d. ACTH
e. cAMP
f. G-protein
g. ATP
h. protein kinase
i. glucocorticoid secretion

___ 32. What mechanism of hormone action does the preceding series of events exemplify?

a. positive feedback mechanism
b. negative feedback mechanism
c. mobile-receptor mechanism
d. fixed-membrane-receptor mechanism
e. all of the above

___ 33. What does the following sequence illustrate? Blood level of thyroxine low > hypothalamus > TRH > adenohypophysis > TSH > thyroid > thyroid hormones > normal hormone level restored > hypothalamus > inhibition of TRH secretion.

a. positive feedback mechanism
b. negative feedback mechanism
c. mobile-receptor mechanism
d. fixed-membrane-receptor mechanism
e. both a and c

___ 34. Where are the receptors for such hormones as insulin and ACTH located?

a. on the outer surface of the target cell
b. on the inner surface of the cell membrane
c. in the cytoplasm of the target cell
d. in the nucleus of the target cell

___ 35. Where are the receptors for such hormones as estrogen and cortisol located?

a. on the outer surface of the target cell
b. on the inner surface of the cell membrane
c. in the cytoplasm of the target cell
d. in the nucleus of the target cell

___ 36. Calcitonin and parathormone have essentially opposite functions in regulating the

a. level of calcium in the blood
b. metabolism of osteoblasts
c. mitotic activity of osteoclasts
d. hardness and density of bone tissue
e. both a and d

___ 37. Parathormone decreases the concentration of phosphate in the blood by

a. promoting its deposition in bone
b. combining it with calcium to form Ca_2PO_4
c. inhibiting its reabsorption by the kidneys
d. promoting its reabsorption by the kidneys
e. converting it to ATP

___ 38. Aldosterone regulates the blood levels of

a. Na^+ and K^+
b. H^+
c. Ca^{2+} and CO_3^-
d. both a and b, above

___ 39. A drop in blood volume causes juxtaglomerular cells in the kidney to release the enzyme

a. angiotensinogen
b. angiotensin I
c. angiotensin II
d. renin
e. both b and c

___ 40. Chromaffin cells are the secretory cells of the

a. adrenal medulla
b. adrenal cortex
c. hypothalamus
d. cardiac endothelium
e. none of the above

Matching

___ 41. cretinism
___ 42. goiter
___ 43. Type I diabetes
___ 44. prolactin-inhibiting hormone
___ 45. inhibin
___ 46. diabetes insipidus
___ 47. melatonin
___ 48. gonadotropic-releasing hormone
___ 49. milk ejection
___ 50. calcitonin

a. adenohypophysis
b. neurohypophysis
c. hypothalamus
d. thyroid
e. pancreas
f. testes
g. pineal gland

Integrative Thinking

1. African pygmies are skilled trackers and intrepid hunters, and they are quite uniformly short of stature. It had been assumed that they are also, by heredity, short of growth hormone. But such does not seem to be the case. Serological tests have shown that their growth hormone titers are within the usual human range. How do you explain this anomaly? Detail your reasoning, considering all the possibilities that you have learned so far.

2. Jan was tested for anabolic steroids immediately after the race. The result was positive, which meant disqualification. Jan pleaded innocence, claiming that a rival had surreptitiously added steroid to the urine sample when nobody was looking. The completed test revealed a very significant amount of steroid in the urine specimen. Does this finding support or refute Jan's claim? Explain. What other signs would the testers look for before rendering a decision if (a) Jan is a woman, or (b) Jan is a man?

3. Melanie is six years old. She takes insulin regularly and her diet is carefully regulated. Her father, Jacob, is diabetic, but he is not on insulin. He is on a very strict diet instead, and he follows a rather demanding exercise program to control his weight. How do you account for the difference between the way father and daughter deal with their diabetes?

Your Turn

This will put your ingenuity to the test. Organize a competition patterned after the popular TV game show, Jeopardy. Let each person in your study group select some hormonally regulated disease or abnormality, of which he or she will present the symptoms or conditions to the rest of the group. Keep score, and the person who correctly identifies the most can introduce and lead a discussion on stress in college life. This could be very important and very revealing to you because (to paraphrase a poetic dictum) "More things are wrought by stress than this world dreams of."

19 The Cardiovascular System: Blood

Active Reading

Introduction

1. Why is blood considered to be a type of connective tissue? _____

2. What are the two components of blood?

    a. _____

    b. _____

I. Functions of Blood (586)

1. What are the three functions of blood?

    a. _____

    b. _____

    c. _____

II. Properties of Blood (586)

1. An average sized (a) woman has ___; (b) man has ___; and (c) newborn has ___

    a. no blood
    b. 4-5 liters of blood
    c. 5-6 liters of blood
    d. 240 mL of blood

___ 2. The viscosity of water is 1.00; the viscosity of blood ranges between

    a. 3.5 and 5.5
    b. 35.00 and 55.00
    c. 1.045 and 1.065
    d. 1.45 and 1.65

____ 3. The specific gravity (density) of blood, as compared to to that of water (1.00), is approximately

    a. 0.950
    b. 1.025
    c. 1.055
    d. 1.085
    e. 1.950

4. What makes arterial blood red?

5. Why is venous blood more acidic than arterial blood?

III. Plasma (586)

1. What percent of blood, by volume, is plasma? _____

Match the following pairs.

____ 2. Leukocytes
____ 3. Thrombocytes
____ 4. Erythrocytes

    a. red blood cells
    b. white blood cells
    c. platelets

Match more pairs.

____ 5. 90% of plasma
____ 6. 7% of plasma
____ 7. 3% of plasma

    a. electrolytes
    b. water
    c. proteins

Match still more pairs.

____ 8. Two forms of lipid transport proteins
____ 9. Swelling caused by fluid leaving bloodstream
____ 10. Foreign body with which antibodies combine
____ 11. Two classes of lipid transport proteins
____ 12. Incomplete plasma that will not clot
____ 13. Antibodies that combine with antigens
____ 14. Water retention proteins
____ 15. Protein essential for clotting

    a. albumins
    b. edema
    c. fibrinogen
    d. serum
    e. alpha and beta globulins
    f. LDLs and HDLs
    g. gamma globulins
    h. antigen

16. List the seven major ions in plasma, and for each identify whether it is an anion or a cation.

    a. _____

    b. _____

    c. _____

    d. _____

    e. _____

    f. _____

    g. _____

17. What is the concentration of glucose in plasma? _____

18. What are the three principal gases dissolved in plasma?

a. _____

b. _____

c. _____

IV. Formed Elements (588)

1. About how many erythrocytes are there in the human body? _____

2. What does a mature erythrocyte lack? _____

3. Identify the components of the illustration in the figure below.

4. How many O_2 molecules may bind to a hemoglobin molecule at any one time? _____

5. Define the following.

a. HbO_2 _____

b. reduced hemoglobin _____

c. HbF _____

6. Why is carbon monoxide fatal in even small amounts? _____

7. What percent of carbon dioxide in the blood takes each of the following forms?

    a. H_2CO_3, HCO_3^-, and H^+ _____

    b. $HbCO_2$ _____

    c. CO_2 _____

8. What does hemolysis result from? _____

9. Where does erythropoiesis occur? _____

10. Arrange the following types of cells in their correct chronological/developmental order: erythrocyte, erythroblast, hemocytoblast, reticulocyte, common myeloid progenitor cell.

    a. _____

    b. _____

    c. _____

    d. _____

    e. _____

11. Where in the body are erythrocytes destroyed? _____

12. What glycoprotein controls the rate of erythropoiesis? _____

13. Describe how the body recycles or disposes of the following elements of a hemoglobin molecule after its erythrocyte has ruptured.

    a. alpha and beta chains _____

    b. heme _____

    c. iron _____

14. Describe the function of the following leukocytes.

    a. neutrophils _____

    b. eosinophils _____

    c. basophils _____

    d. monocytes _____

    e. lymphocytes _____

15. What are the two classes of leukocytes?

    a. _____

    b. _____

16. Which are the largest blood cells? _____

17. How long do monocytes remain in the bloodstream? _____

18. What do monocytes eventually become? _____

19. What areas of the body do lymphocytes call home?

    a. _____

    b. _____

    c. _____

    d. _____

    e. _____

    f. _____

20. True or false: While all erythrocytes contain hemoglobin, and are necessary for the transport of oxygen, all leukocytes are phagocytes, and crucial for the functioning of the immune system. ___

21. What are the two types of lymphocytes?

    a. _____

    b. _____

22. What are plasma cells, and where do they come from? _____

23. What is the main function of platelets? _____

24. From what cells do platelets derive? _____

25. To what do platelets adhere? _____

26. To what do platelets not adhere? _____

27. What is hemostasis? _____

V. Clinical Blood Tests (596)

1. What three blood tests are routinely conducted?

    a. _____

    b. _____

    c. _____

2. What is a normal RBC count for women? _____. For men? _____

3. What is the normal HCT for men? _____. For women? _____

___ 4. An HCT less than 30 indicates

    a. an abnormally low RBC count
    b. anemia
    c. every 100 mL of whole blood contains less than 30 mL of red blood cells
    d. all of the above

5. What does a hemoglobinometer measure? _____

   Describe how it works. _____

6. Indicate the expected results of a differential white count under the following conditions.

    a. various types of leukemias _____

    b. allergic reaction _____

    c. radiation exposure _____

    d. stress, inflammation, or bacterial infection _____

    e. viral infection _____

    f. immunosuppressive agent present _____

VI. Hemostasis: The Prevention of Blood Loss (597)

1. What are the three elements of the hemostatic mechanism?

    a. _____

    b. _____

    c. _____

Match the following.

____ 2. Smooth muscle tissue contracts

____ 3. Blood clotting occurs if platelets alone are unable to halt bleeding

____ 4. Vascular compression caused by pressure of lost blood in surrounding tissue

____ 5. Endothelial cells secrete VWF

____ 6. Hemostatic plug clogs small opening in vessel walls

a. vasoconstrictive phase
b. platelet phase
c. coagulation phase

7. Complete the following table.

	Factor Name	Function	Description
I			
II			
III			
IV			
V			
VI			
VII			
VIII			
IX			
X			
XI			
XII			
XIII			

8. Describe the events in each of the five stages of the coagulation phase.

a. _____

b. _____

c. _____

d. _____

e. _____

9. In the figure below, identify examples of the following: platelets, erythrocytes, fibrin.

10. With which pathway are the clotting factors associated? _____

11. What triggers the extrinsic pathway? _____

12. At what point do the extrinsic and intrinsic pathways merge? _____

13. Which pathway is faster? _____

14. Describe the cascade effect. _____

15. Describe the causes and effects of hypovolemic shock. _____

_____

VII. Anticoagulation: The Inhibition of Clotting (602)

1. Why is it important that blood not clot unnecessarily? _____

2. In what two ways do the blood vessels inhibit clotting?

a. _____

b. _____

3. What type of molecule is heparin? _____

4. How does antithrombin-heparin prevent clotting? ______

5. Describe how fibrinolysis takes place. ______

6. What is the best known and most commonly used anticoagulant drug? ______

Match the following pairs.

___ 7. Normal coagulation takes place at this or higher
___ 8. Normal prothrombin time
___ 9. Normal bleeding time in fingertip or ear lobe
___ 10. Normal clotting time

a. 5 to 8 minutes
b. 150,000 per cubic millimeter
c. 3 to 6 minutes
d. 12 seconds

VIII. Blood Types (603)

1. Define the following.
   a. agglutinogens ______
   b. agglutination ______
   c. hemaglutination ______

Match the following pairs.

___ 2. Type A
___ 3. Type B
___ 4. Type AB
___ 5. Type O

a. agglutinogen A
b. agglutinogen B
c. neither agglutinogen
d. both agglutinogens

6. Why should blood be cross-matched prior to transfusion? ______

7. Describe why it is potentially dangerous for an Rh negative woman to conceive an Rh positive fetus. ______

IX. When Things Go Wrong (606)

Match the following pairs.

___ 1. Infecting organism enters red blood cells, reproducing until the cells burst
___ 2. Inherited hemoglobin S producing gene causes amino acid substitution in two of four protein chains in the hemoglobin molecule
___ 3. Number of red blood cells, normal concentration of hemoglobin, or hematocrit below normal
___ 4. Improper absorption of dietary vitamin B_{12}
___ 5. Failure of bone marrow to function properly
___ 6. Iron is not available to make erythrocytes
___ 7. Results from heavy blood loss

a. anemia
b. hemorrhagic anemia
c. iron-deficiency anemia
d. aplastic anemia
e. hemolytic anemia
f. pernicious anemia
g. sickle cell anemia

8. True or false? Hemophilia is carried by females, but expressed most often in males. ____

9. Name four varieties of leukemia.

     a. _____

     b. _____

     c. _____

     d. _____

Key Terms

ABO blood grouping 603	erythropoiesis 590	leukocyte 591
agglutination 603	extrinsic pathway 600	lymphocyte 594
agglutinin 603	fibrin 598	megakaryocyte 595
agglutinogen 603	fibrinogen 587, 598	monocyte 594
aggregation 598	formed elements 585, 588	neutrophil 594
agranulocyte 594	granulocyte 594	plasma 585, 586
basophil 594	heme 589	platelet 595
coagulation 596	hemoglobin 588-596	Rh factor 604
common pathway 600	hemostasis 597	thrombin 598
eosinophil 594	heparin 602	thrombocyte 595
erythrocyte 588	intrinsic pathway 600	tPA 602

Post Test

Multiple Choice

____ 1. The functions of blood include, among other things

     a. the transport of O_2, nutrients, hormones, and metabolic wastes

     b. the regulation of body temperature, pH and amount of bodily fluids, and electrolyte levels

     c. protecting the body against infectious diseases

     d. maintaining homeostasis by a, b, and c

     e. all of the above

____ 2. The pH of arterial blood usually ranges between

     a. 6.8 and 7.0

     b. 6.9 and 7.2

     c. 7.0 and 7.42

     d. 7.35 and 7.45

     e. 7.45 and 7.55

___ 3. Whole blood consists of two parts

    a. blood cells and plasma
    b. formed elements and serum
    c. plasma and serum
    d. red blood cells and white blood cells
    e. red blood cells and plasma cells

___ 4. The albumins, fibrinogen, and the globulins constitute the major

    a. chemical components of blood
    b. formed elements of blood
    c. blood electrolytes
    d. blood proteins
    e. causes of edema

___ 5. The most abundant among the components named in the preceding question is/are

    a. the albumins
    b. fibrinogen
    c. the alpha globulins
    d. the beta globulins
    e. the immunoglobulins

Matching

___ 6. Neutrophils
___ 7. Eosinophils
___ 8. Red blood cells
___ 9. Produced by megakaryocyte
___ 10. Some contain heparin and histamine
___ 11. Include the largest blood cell
___ 12. Derived from bone marrow myeloblasts
___ 13. Some derived from bone marrow monoblasts
___ 14. Function in blood clotting
___ 15. Firmly bond carbon monoxide
___ 16. Include the lymphocytes
___ 17. Strongly phagocytic
___ 18. Enucleate cell fragments
___ 19. Contain large amounts of carbonic anhydrase
___ 20. Include B cells and T cells

    a. erythrocytes
    b. granulocytes
    c. agranulocytes
    d. platelets

___ 21. Swelling caused by fluid leaving bloodstream
___ 22. Function in lipid transport
___ 23. Incomplete plasma that will not clot
___ 24. Transport cholesterol
___ 25. Antibodies
___ 26. Antibody targets
___ 27. Essential to clotting process
___ 28. Promote water retention in the blood

    a. albumins
    b. fibrinogen
    c. alpha and beta globulins
    d. gamma globulins
    e. serum
    f. HDLs and LDLs
    g. edema
    h. antigens

Matching

___ 29. Destroy and digest invading microorganisms; contain plasminogen;
cytoplasmic granules stain with acid
dyes

___ 30. Mobile phagocytes with large,
folded nucleus, and frequently
containing fine cytoplasmic granules

___ 31. Have cytoplasmic granules containing slow-reacting-substance A
(SRS-A)

___ 32. Small, agranular cells with single
round nucleus; concentrated in
spleen, tonsils, adenoids, thymus,
and nodular aggregations

___ 33. Most abundant phagocytic cells, rich
in lysosomes

___ 34. Form from enlarged, activated B
cells; efficient antibody producers

a. neutrophils
b. eosinophils
c. basophils
d. monocytes
e. lymphocytes
f. plasma cells

___ 35. Determines the percentage of the
various leukocytes.

___ 36. Determines the number of erythrocytes per mm^3 of blood.

___ 37. Estimates the number of leukocytes
per mm^3 of blood.

___ 38. Measures volume percent red blood
cells removed from whole blood by
centrifugation.

___ 39. Determines by photometry the
concentration of Hb in a sample of
whole blood.

a. hemocytometer
b. hematocrit
c. hemoglobinometer
d. differential count
e. white blood cell count

___ 40. Indicative of certain viral infections.
___ 41. May result from stress, inflammation, or bacterial infection.
___ 42. Characterizes some forms of leukemia.
___ 43. Often characteristic of allergic
reactions.
___ 44. Frequently caused by exposure to
ionizing radiation or certain drugs.
___ 45. Caused by immunosuppressive
agents, such as the cortical steroids.

a. high neutrophil count
b. low neutrophil count
c. high lymphocyte count
d. low lymphocyte count
e. high eosinophil count
f. high basophil, lymphocyte, or monocyte
count

___ 46. Has anti-A antibodies in plasma
___ 47. Has anti-B antibodies in plasma
___ 48. Has both anti-A and anti-B antibodies in plasma
___ 49. Has neither anti-A nor anti-B antibodies in plasma
___ 50. The so-called "universal donor"

a. blood Type A
b. blood Type B
c. blood Type AB
d. blood Type O

___ 51. Thalassemia
___ 52. Due to vitamin B_{12} or intrinsic factor deficiency
___ 53. Sickle cell anemia
___ 54. Leads to hypochromic, microcytic anemia
___ 55. Due to dysfunctional bone marrow caused by poisoning or excessive exposure to ionizing radiation
___ 56. Caused by heavy loss of blood

a. hemorrhagic anemia
b. iron-deficiency anemia
c. aplastic anemia
d. hemolytic anemia
e. pernicious anemia

57. Indicate the sequence of the reactions that lead to the generation of a blood clot.
(__) → (__) → (__) → (__)

a. fibrinogen + thrombin
b. thromboplastinogen + thromboplastinogenase + antihemophilic factor
c. prothrombin + thromboplastin + Ca^{2+}
d. fibrin

Integrative Thinking

1. Imagine a person who has always lived in a coastal Peruvian village, then moved to a village high in the Andes. What physiological problems does this person experience immediately, and what mechanisms come into play to adapt this person to the change in altitude?

2. Mrs. O'Grady was quite alarmed at first sight of her newborn baby. Its skin was yellow! Explain what might be wrong, and what circumstances could have led to this condition.

3. A 42-year-old man develops leukemia. His neighbor's 6-year-old daughter also develops leukemia. They live close to a petrochemical factory that has been cited by the Environmental Protection Agency for a series of violations. As a cub reporter, you are assigned the job of covering this story. Outline what you would write that would be informative, objective, and interesting.

4. Why is bone marrow transplantation considered a hazardous procedure?

Your Turn

Arrange with your instructor to borrow a stethoscope (perhaps two, if your group numbers more than four or five members). Learn how to measure pulse and blood pressure on each other, and then run a few simple experiments, recording the values that you determine. Suggestion for experiments:

1. Subject at rest: pulse rate, blood pressure

2. Subject after a few pushups or running in place (no need to be heroic; four or five pushups and/or 10 seconds of running in place should be enough): p.r., b.p.

3. Subject at intervals after exercise: 20 seconds, 1 minute, 3 minutes

4. Determine a new baseline at rest, followed by having the subject try to do something under mental or emotional pressure, such as multiplying several numbers rapidly, naming the capitals of states or countries as quickly as possible after having the states or countries named (Hungary! Quick!! Australia! Hurry up! Delaware! Come on! Faster!! What's your boyfriend's (girlfriend's) name? What's your favorite food? Who won the American League pennant last year? What's 43 times 12? Hurry up!). Then measure p.r. and b.p. Compare the effects of physical exercise and mental or emotional pressure.

20 The Cardiovascular System: The Heart

Active Reading

Introduction

Match the following.

____ 1. Movement of blood through the limbs

____ 2. Movement of blood through the digestive tract

____ 3. Movement of blood through the brain

____ 4. Movement of blood through the lungs

a. systemic circulation
b. pulmonary circulation

5. Why is the heart a double pump? _____

____ 6. Blood enters the heart through the

a. atria
b. ventricles
c. arteries
d. capillaries
e. foramen ovale

____ 7. Blood leaves the heart through the

a. atria
b. ventricles
c. veins
d. foramen ovale

8. Although a well-tuned athlete might have a heart rate between 30 and 50 beats per minute, what is a more usual, less fit heart rate? _____

I. Anatomy of the Heart (614)

1. What is the weight of an average female heart? _____

2. What is the weight of an average male heart? _____

3. On the figure below, identify the following anatomical structures: apex, base, aortic arch, pulmonary trunk, diaphragm, thoracic aorta.

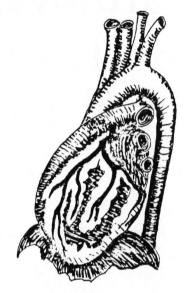

Left-side View

4. Describe the role of gap junctions and desmosomes in the coordination of cardiac muscle cell contractions. _____

5. On the figure below, identify the following: wall of heart, pericardium, endocardium, myocardium, epicardium, pericardial cavity, fibrous pericardium, parietal layer of serous pericardium.

6. Of what significance are numerous mitochondria and an abundant supply of myoglobin to the myocardium? _____

7. What are the five elements of the cardiac skeleton?

    a. _____

    b. _____

    c. _____

    d. _____

Match the following pairs.

  ___ 8. Location of fetal foramen

  ___ 9. Indication of border between atria and ventricles

  ___ 10. Papillary muscles and supportive cords attached to the free edges of the atrioventricular valves

  ___ 11. Upper part of right ventricular wall

  ___ 12. Ridges on auricles and anterior wall of right atrium

  ___ 13. Coarse bundles of cardiac muscle in both ventricles

  a. coronary sulcus
  b. infundibulum
  c. fossa ovalis
  d. musculi pectinati
  e. chordae tendineae
  f. trabeculae carneae

14. What are the four valves of the heart?

    a. _____

    b. _____

    c. _____

    d. _____

15. Which of the valves is/are bicuspid? _____

16. Which of the valves is/are tricuspid? _____

17. What is the structural difference between the AV valves and the semilunar valves? ____

_____

18. Describe the structure and/or function of the following.

    a. chordae tendineae _____

    b. pulmonary arteries _____

    c. pulmonary veins_____

    d. superior vena cava _____

    e. inferior vena cava_____

    f. aorta_____

    g. coronary circulation _____

19. On the figure below, identify the following: coronary sinus, right coronary artery, left coronary artery, circumflex branch of left coronary artery, anterior interventricular branch of left coronary artery, left anterior descending branch of coronary artery, posterior interventricular branch of right coronary artery, small cardiac vein.

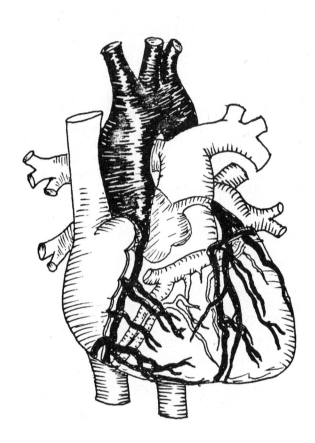

20. Identify the regulating mechanism of blood flow in each of the following.

    a. metabolic process of autoregulation _____

    b. myogenic process of autoregulation _____

21. What is the correct sequence of the following events, beginning with a?

    a → (__) → (__) → (__)

    a. The ventricles contract, sending blood to the body and lungs
    b. Blood is pumped into the ventricles
    c. Blood enters the atria
    d. The ventricles, filled with blood, relax during the AV nodal delay

II. Physiology of the Heart (627)

1. What properties of the heart enable it to beat reliably for a lifetime? _____

_____

2. Complete the following table.

Phase	Electrical Potential	Ion Movement	Duration
Depolarization			
Early repolarization			
Plateau			
Repolarization			
Resting potential			

3. What is the function of pacemaker cells? _____

4. On the figure below, identify the following: superior vena cava, AV node, SA node, Purkinje fibers, right and left bundle branches, AV bundle, internodal tracts, interatrial tract.

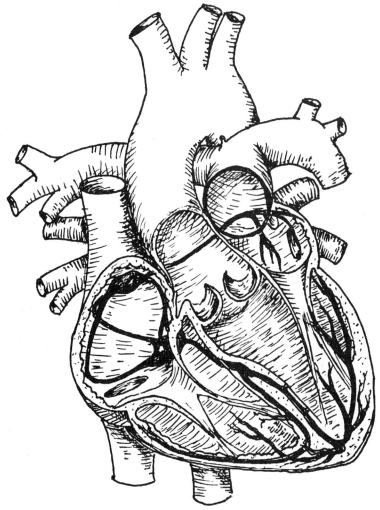

5. List the proper sequence by which an impulse travels through the tissues identified in the previous question.

   a. _____

   b. _____

   c. _____

   d. _____

   e. _____

6. What is the heart's normal pacemaker? _____

7. What is the heart's secondary pacemaker? _____

8. What are the heart's tertiary pacemakers? _____

9. Why are the latter pacemakers inadequate? _____

10. Complete the following table.

ECG Event	Duration	Physiological Event
P wave		
		Ventricular depolarization plus ventricular repolarization
T wave		
		Atrial depolarization and conduction through AV node
P-R interval		
		Ventricular depolarization masking repolarization of atria
S-T interval		
		End of ventricular depolarization to beginning of ventricular repolarization

11. How long does it take the heart to pump blood in a complete cycle through the body? __

_____

12. What are four distinct physiological events associated with a single heartbeat or cardiac cycle?

   a. _____

   b. _____

   c. _____

   d. _____

13. Describe the four stages and their duration in a cardiac cycle.

    a. _____

    b. _____

    c. _____

    d. _____

14. On the figure below, label each of the waves or spikes, and specify what each represents.

    P _____

    Q _____

    R _____

    S _____

    T _____

15. Define the following two terms.

    a. auscultatory event _____

    b. phonocardiogram _____

16. Where on the heart can one best hear the lubb with a stethoscope? _____

17. What event does the dupp announce? _____

Match the following pairs.

____ 18. Quantity of blood pumped by either   a. cardiac index
         ventricle in 1 minute   b. cardiac reserve
____ 19. Quantity of blood pumped by either   c. cardiac output
         ventricle in 1 minute divided by
         body surface area
____ 20. Difference, as a percentage, between
         the amount of blood the heart can
         pump under stressful conditions and
         the amount it pumps during restful
         conditions

21. What is the effect on the pacemaker rate of the following?

    a. the sympathetic nervous system _____

    b. the parasympathetic nervous system _____

    c. acetylcholine _____

    d. norepinephrine _____

22. What is the effect on stroke volume caused by an increase in each of the following?

    a. preload _____

    b. contractility _____

    c. afterload _____

23. What is the Frank-Starling law? _____

24. Why will a heart continue to beat even after it is severed from the nervous system?

    _____

25. Where is the main neural control center for the heart? _____

___ 26. Chemoreceptors

    a. measure changes in blood pressure
    b. measure changes in heart rate
    c. measure changes in the gases and other substances dissolved in blood
    d. none of the above

___ 27. Baroreceptors

    a. measure changes in blood pressure
    b. measure how much arteries are stretched
    c. relay information to the medulla oblongata
    d. all of the above

28. Describe the role of the adrenal medulla in regulating cardiac output. _____

    _____

III. The Effects of Aging on the Heart (644)

1. What is the effect of age on the maximum heart rate? _____

2. How does a hardening of the arteries affect the heart? _____

IV. Developmental Anatomy of the Heart (644)

1. When during development do each of the following form?

    a. heart cords _____

    b. heart tubes _____

    c. endocardial heart tube _____

    d. bulbus cordis _____

    e. first ventricle _____

    f. first atrium _____

g. pacemaker _____

h. interatrial septum _____

i. interventricular septum _____

j. closing of foramen ovale _____

V. When Things Go Wrong (646)

Match the following pairs.

____ 1. Myocardial infarction
____ 2. Congestive heart failure
____ 3. Ventricular septal defect
____ 4. Interatrial septal defect
____ 5. Tetralogy of Fallot

a. combination of four congenital heart diseases
b. opening in the ventricular septum
c. heart attack caused by death of myocardial tissue
d. failure of the foramen ovale to close at birth
e. ventricles do not pump equal amounts of blood, or pump less blood than enters the atrium

6. What is a possible correction for valvular heart disease? _____

7. List four distinct infectious heart diseases.

a. _____

b. _____

c. _____

d. _____

8. At what point does the body experience hypovolemic shock? _____

9. What is the usual cause of cardiogenic shock? _____

10. What condition indicates a rapid rise in pressure within the pericardial sac? _____

11. List five cardiac arrhythmias.

a. _____

b. _____

c. _____

d. _____

e. _____

12. What relationship does excess iron have to heart disease? _____

13. What causes angina pectoris? _____

Key Terms

aorta 621
atrioventricular bundle 631
atrioventricular node 631
atrioventricular valves 621
atrium 619
bicuspid valve 621
cardiac conducting myofibers 631
cardiac cycle 633
cardiac skeleon 617
cardioregulatory center 641
coronary circulation 621
coronary sinus 624
diastole 633

electrocardiogram (ECG) 631
endocardium 617
epicardium 616
Frank-Starling law 640
inferior vena cava 621
mitral valve 621
myocardium 616
P wave 633
pacemaker cells 629
pericardium 614
pulmonary arteries 621
pulmonary circulation 614
pulmonary veins 621

QRS complex 633, 635
semilunar valves 621
septum 619
sinoatrial node 629
stroke volume 635, 640
superior vena cava 621
systemic circulation 614
systole 633
T wave 635
tricuspid valve 621
ventricle 619

Post Test

Multiple Choice

____ 1. The human blood circulatory system is

    a. a network of blindly ending tubes and vessels
    b. an open system of tubes and vessels
    c. a closed system of tubes and vessels
    d. partly open and partly closed

____ 2. Cardiac muscle is found

    a. only in the heart
    b. only in the heart and the aorta
    c. primarily in the heart and the great veins
    d. in the heart and pulmonary arteries

____ 3. In the normal standing position, the uppermost part of one's heart is called the

    a. ventricles
    b. sinus venosus
    c. apex
    d. base

____ 4. The thickest part of the heart is the wall of the

    a. right ventricle
    b. left ventricle
    c. inter atrial septum
    d. inter ventricular septum

____ 5. The cells of cardiac muscle are connected end-to-end by

    a. a syncytium
    b. tight junctions
    c. intercalated disks
    d. myoglobin

____ 6. Blood is prevented from reversing
its flow in the heart by the

    a. cardiac skeleton
    b. chordae tendineae
    c. papillary muscles
    d. all of the above

____ 7. The left AV valve is known as the

    a. mitral valve
    b. tricuspid valve
    c. semilunar valve
    d. AV orifice

____ 8. The valve that lies in the opening
leading out of the right ventricle is
the

    a. mitral valve
    b. tricuspid valve
    c. aortic semilunar valve
    d. pulmonary semilunar valve

____ 9. Blood enters the right atrium from
the

    a. pulmonary veins
    b. superior and inferior venae cavae
    c. ascending and descending aortae
    d. all of the above

____ 10. The heart muscle is nourished by
blood passing through

    a. atria and ventricles
    b. the pulmonary artery
    c. the aorta
    d. the coronary circulation

____ 11. Blood flow to the heart muscle is
regulated by the

    a. metabolic and myogenic processes
    b. adrenal medulla
    c. coronary endothelium
    d. coronary sinus
    e. both c and d

____ 12. The direction of blood flow through
the heart chambers and intervening
circuit is, in order

    a. rt. atrium > rt. ventricle > pulmonary
circuit > left ventricle > left atrium
    b. rt. atrium > rt. ventricle > systemic circ. >
left atrium > left ventricle
    c. left atrium > left ventricle > pulmonary
circ. > rt. atrium > rt. ventricle
    d. rt. atrium > rt. ventricle > systemic circ. >
left atrium > left ventricle
    e. rt. atrium > rt. ventricle > pulmonary circ.
> left atrium > left ventricle

Matching

_____ 13. Located in the interventricular septum

_____ 14. Located in the wall of the right atrium

_____ 15. Located in the ventricular muscle

_____ 16. Convey impulses between SA and AV nodes

_____ 17. Has inherent rythmicity of 40-50 beats/min.

_____ 18. Has inherent rythmicity of 70-80 beats/min.

_____ 19. Has inherent rythmicity of 20-40 beats/min.

_____ 20. Develops from specialized region of embryonic sinus venosus

_____ 21. Aids spread of depolarization to wall of left atrium

a. SA node
b. interatrial tracts
c. internodal tracts
d. AV node
e. AV bundle
f. Purkinje fibers

Matching

_____ 22. Ca^{2+} moves into ventricular muscle cells through slow voltage-gated channels.

_____ 23. Repolarization momentarily levels off short of its normal potential of -90 mV.

_____ 24. Na^+ moves into muscle cells through fast voltage-gated channels.

_____ 25. Permeability to Na^+ increases to change membrane potential from -90 mV to $+30$ mV.

_____ 26. K^+ channels open, Ca^{2+} channels close, membrane potential restored to -90mV.

_____ 27. Resting membrane potential is maintained, but as an unstable pacemaker potential.

a. depolarization
b. early repolarization
c. plateau phase
d. repolarization
e. active transport

On the figure below, identify each of the ECG components indicated by a through e.

___ 28. T wave
___ 29. P wave
___ 30. ST segment
___ 31. PR segment
___ 32. QRS complex

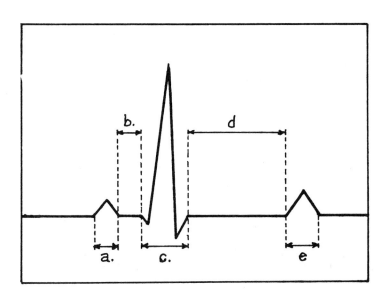

Match the relationship between the ECG components identified in the preceding question and the following events.

___ 33. Depolarization of ventricles
___ 34. Repolarization of ventricles
___ 35. Atrial depolarization preceding atrial systole
___ 36. Atrial depolarization and conduction through AV node
___ 37. End of ventricular depolarization to beginning of repolarization of the ventricles

Matching

___ 38. Auscultatory event
___ 39. Possible valvular dysfunction
___ 40. Start of ventricular systole
___ 41. Recoil from closing of semilunar valves
___ 42. Heard in cases of congenital heart disease
___ 43. Gentle sound between first and second heart sounds

a. lubb
b. dupp
c. resting heart murmur
d. lubb-hiss-dupp, lubb-hiss-dupp
e. all of the above

Matching

____ 44. mL/beat
____ 45. mL/beat X beats/min.
____ 46. CO ÷ body surface area in meters²
____ 47. The more cardiac muscle is
         stretched, the stronger its contraction
         will be
____ 48. The volume of blood pumped under
         stressful conditions minus the
         volume pumped under normal
         conditions
____ 49. Mechanisms that strongly influence
         venous return
____ 50. Directly increased by positive
         inotropic agents, such as calcium,
         epinephrine, and glucagon

a. cardiac output
b. cardiac index
c. cardiac reserve
d. stroke volume
e. Frank-Starling law
f. skeletal muscle and respiratory pumps

Matching

____ 51. Mitral insufficiency
____ 52. Inflammation of the heart muscle
         per se
____ 53. Pulmonary edema, for example
____ 54. Backflow into right ventricle during
         diastole
____ 55. Morbidity (dying or death) of an
         area of heart tissue

a. myocardial infarction
b. congestive heart failure
c. valvular heart disease
d. pulmonary stenosis
e. myocarditis

Matching

____ 56. Infection of the lining epithelium of
         the heart and valves
____ 57. Referred pain owing to damaged or
         blocked coronary artery
____ 58. Weak, uncoordinated, ineffective
         fluttering
____ 59. dysfunction of AV node due to
         disease
____ 60. consequence of greatly reduced
         cardiac output

a. endocarditis
b. circulatory shock
c. fibrillation
d. AV block
e. angina pectoris

Integrative Thinking

1. A patient enters the emergency ward complaining of shortness of breath and pains in his chest. Naturally, he is quite alarmed. Preliminary tests point to the likelihood of an obstruction in the coronary circulation. Describe three tests by which this diagnosis can be confirmed, and two ways of treating such an obstruction.

2. An ECG shows the tracing below. The pulse rate, although a bit irregular, is about 45 beats per minute. What is a likely interpretation?

3. Instead of having two pacemaker nodes, SA and AV, in the heart, why wouldn't just one suffice, if it were located at the center of the heart, where the atria and ventricles meet?

Your Turn

Time for discussion! In anticipation of the next meeting of your study group, let each member prepare a list of the three or four concepts or processes described in this chapter that he or she has had difficulty in understanding. Present these problems to the group for discussion. John may have a problem with topic A, but Joan has grasped it clearly. Joan, on the other hand, cannot understand topic B, but John does, and so on. Finally, if there are any problems that have not been resolved in this way, make a list of them and present them to your instructor.

21 The Cardiovascular System: Blood Vessels

Active Reading

Introduction

1. What makes up the circulatory system, as it is referred to in this chapter? _____

_____

I. Types of Blood Vessels (654)

Match the following pairs.

____	1. Carry blood to veins	a. elastic arteries
____	2. Walls expand with each heartbeat	b. muscular arteries
____	3. Branch into arterioles	c. arterioles
____	4. Carry blood to heart	d. metarterioles
____	5. Porous walls allow passage of water and small particles	e. capillaries
____		f. venules
____	6. Emerge from arterioles, lead into capillary network	g. veins
____	7. Branches of muscular arteries	

8. Why, in illustrations, are arteries generally colored red, and veins blue? _____

_____

9. What are the two major blood vessel trunks?

a. _____

b. _____

10. Complete the following table.

Name	Location in Artery	Composed of
Tunica intima		
	Central canal	N.A.
		Collagen fibers, elastic fibers, nerves, and lymphatic vessels
	Middle covering of arteries	

11. What vessels nourish the larger blood vessels? _____

12. What layer makes up the walls of capillaries? _____

13. True or false? All capillaries have about the same diameter. ___

14. Is cross-sectional diameter a good criterion for classifying blood vessels? _____

15. Name the three types of capillaries, and describe where in the body each type is found.

   a. _____

   b. _____

   c. _____

16. How long does blood remain in a typical capillary? _____

17. Describe the process of autoregulation at the precapillary sphincter. _____

   _____

18. What type of blood flows through the systemic veins? _____

19. Which veins carry oxygenated blood? _____

20. Where do you find the sequence: capillaries → vein → capillaries → vein? _____

21. Describe how the walls of veins differ from those of arteries. _____

   _____

22. What prevents blood from flowing backwards in larger veins? _____

II. Pulmonary and Systemic Circulation of the Blood (660)

1. How long does pulmonary circulation take? _____

2. How long does systemic circulation take? _____

3. Into what atrium do the pulmonary veins flow? _____

4. From which ventricle does the systemic circulation flow? _____

5. Name the four divisions of the systemic circulation.

    a. _____

    b. _____

    c. _____

    d. _____

6. Where do the following two systems carry blood?

    a. hypothalamic-hypophyseal portal system _____

    b. hepatic portal system _____

7. What advantage is there in having the sinusoids of the liver receive blood from abdominal veins? _____

8. What four arteries transport blood to the brain?

    a. _____

    b. _____

    c. _____

    d. _____

Match the following pairs.

___ 9. Supply medial parts of cerebral hemispheres
___ 10. Located on ventral surface of brainstem
___ 11. Supply lateral parts of cerebral hemispheres
___ 12. Supply the eyes
___ 13. Receives blood from basilar artery

a. basilar artery
b. posterior cerebral arteries
c. ophthalmic arteries
d. anterior cerebral arteries
e. middle cerebral arteries

14. Describe how the cutaneous circulation is involved in heat regulation of the body.

    _____

15. What are two systems for regulating the amount of blood reaching the skeletal muscles?

    a. _____

    b. _____

16. Why must fetal circulation differ from adult circulation?

    a. _____

    b. _____

17. What are two mechanisms by which the fetal circulatory system, to a large extent, bypasses the lungs?

    a. _____

    b. _____

18. True or false? The blood of the fetus mixes with that of the mother in the placenta. ___

19. True or false? The umbilical vein carries oxygen-poor blood back to the placenta. ___

20. True or false? The umbilical vein carries oxygen-poor blood from the placenta to the fetus. ___

21. True or false? Fetal blood has a lower oxygen content than adult blood. ___

22. Describe the mechanism by which the foramen ovale closes. _____

    _____

23. What happens to the ductus arteriosis? _____

III. Major Arteries and Veins: Regions Supplied and Drained (665)

1. What two arterial trunks emerge from the heart?

    a. _____

    b. _____

2. What four large veins drain into the heart?

    a. _____

    b. _____

    c. _____

    d. _____

3. What are the two systems for naming arteries and veins?

    a. _____

    b. _____

4. On the figure below, identify the following: ascending aorta, aortic arch, thoracic aorta, abdominal aorta, common carotid artery, vertebral artery, subclavian artery, coronary artery, posterior intercostal arteries, median sacral artery, right common iliac artery.

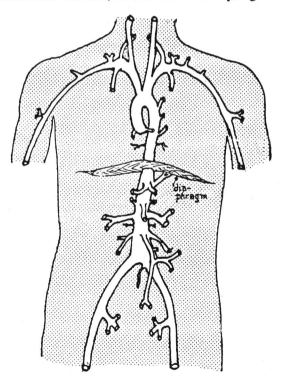

5. What region of the body do the following arteries serve?

    a.   superior phrenic _____

    b.   inferior phrenic _____

    c.   celiac _____

    d.   superior mesenteric _____

    e.   inferior mesenteric _____

    f.   median sacral _____

6. On the figure below, identify the following: middle cerebral artery, anterior cerebral artery, ophthalmic artery, basilar artery, subclavian artery, external carotid artery, internal carotid artery, common carotid artery, vertebral artery, superior thyroid artery.

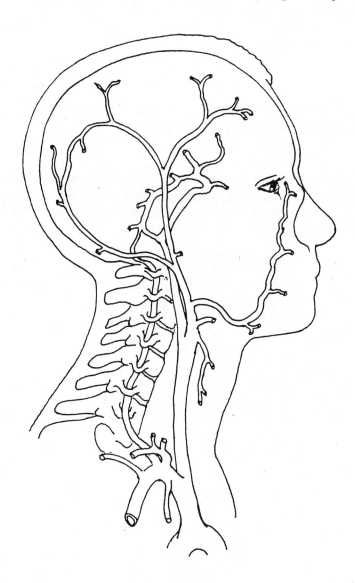

7. On the figure below, identify the following: thyrocervical trunk, axillary artery, brachial artery, radial artery, ulnar artery, digital arteries, subclavian artery.

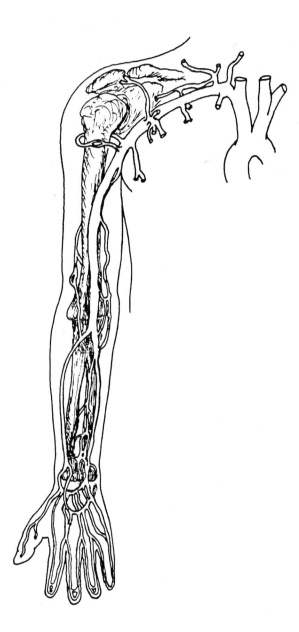

8. Which artery is the most common site for measurement of blood pressure? _____

9. On the figure below, identify the following: common iliac artery, internal iliac artery, external iliac artery, femoral artery, popliteal artery, anterior tibial artery, posterior tibial artery, medial plantar artery.

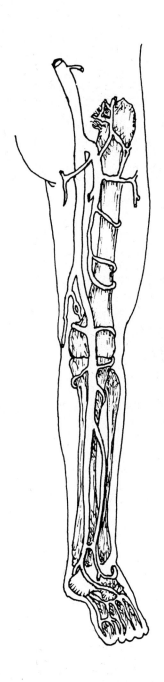

10. On the figure below, identify the following: facial vein, thyroid vein, external jugular vein, vertebral vein, internal jugular vein, sigmoid sinus.

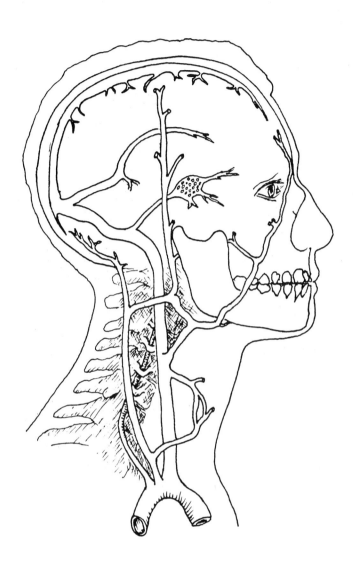

_____ 11. How attractive is text Figure 21.17B?   a. very attractive
                                                    b. mildly attractive
                                                    c. unattractive
                                                    d. very unattractive
                                                    e. thank goodness for the integumentary
                                                       system

12. Into what vein does the external jugular vein drain? _____

13. What vein is often used for withdrawal of blood? _____

15. On the figure below, identify the following: common iliac vein, internal iliac vein, external iliac vein, femoral vein, great saphenous vein, anterior tibial vein, dorsal pedal vein, dorsal venous arch.

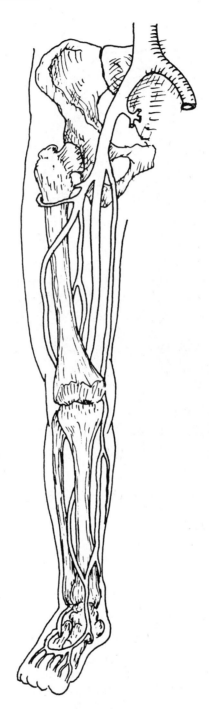

16. What class of veins in the lower limb is organized as venae comitantes? _____

17. Into what does the hemiazygous vein drain? _____

IV. Physiology of the Circulatory System (684)

1. Through which vessels does the body regulate the circulatory system in order to maintain proper pressure and flow? _____

2. What is hemodynamics? _____

3. True or false? Blood flow is pulsatile because of the elasticity of arteries near the heart. ____

4. True or false? Blood flow is least pulsatile and turbulent in the arteries. ____

5. True or false? Blood flow is usually measured in millimeters per minute. ____

6. True or false? Blood pressure, like the barometric pressure of air, is measured in millimeters of mercury (mmHg). ____

7. True or false? Blood pressure is directly proportional to blood velocity, but inversely proportional to resistance. ____

8. Where is blood velocity the fastest?

9. Where is blood velocity the slowest?

10. True or false? Blood velocity is directly proportional to the cross-sectional area of the vessels. ____

11. True or false? Total peripheral resistance represents the friction generated by the walls of all the blood vessels. ____

12. What are units of resistance? _____

13. On what four factors does resistance depend?

    a. _____

    b. _____

    c. _____

    d. _____

14. Where does the energy lost to resistance go (how is it dissipated)? _____

15. Fill in the blanks in the equations below.

    a. $P = F \times (\underline{\quad})$

    b. $F = \Delta P \div (\underline{\quad})$

    c. $R = (\text{length} \times \text{viscosity}) \div (\underline{\hspace{2cm}})$

16. What is the most important factor regulating blood flow? _____

17. What is vasomotion? _____

18. What is the effect of epinephrine and norepinephrine on blood flow? _____

19. What is the effect of vasopressin, oxytocin, and angiotensin II on blood flow? _____

20. What are the effects of the following hormones on blood flow?

    a. histamine _____

    b. atriopeptin _____

    c. bradykinin _____

21. What metabolic factors can cause vasodilation via the process of autoregulation?

    a. _____

    b. _____

    c. _____

    d. _____

    e. _____

22. What metabolic factors can cause vasoconstriction via the process of autoregulation?

    a. _____

    b. _____

23. What primarily determines the pressure within a vessel when the ventricles contract?

    _____

24. If cardiac output increases and the volume of the blood vessels remains the same, what happens to blood pressure? _____

25. Where is the cardioregulatory center? _____

26. What do baroreceptors monitor? _____

27. Where are chemoreceptors located? _____

28. What is the main function of chemoreceptors? _____

29. What is the effect of fear or rage on blood pressure? _____

30. What is the effect of diminished blood viscosity on blood pressure? _____

31. In what two ways do the kidneys regulate blood pressure?

    a. _____

    b. _____

32. What are the four mechanisms that can move materials out through the capillary walls?

    a. _____

    b. _____

    c. _____

    d. _____

33. What are the two opposing forces, the difference between which determines the rate and direction of filtration?

    a. _____

    b. _____

34. How is fluid that is not reabsorbed by the capillaries ultimately returned to the circulatory system? _____

35. What are the five major factors influencing the return of venous blood?

    a. _____

    b. _____

c. _____

d. _____

e. _____

36. The pressure gradient for venous return is the difference in blood pressure between what two locations?

    a. _____

    b. _____

37. Describe the mechanism of the skeletal muscle venous pump. _____

    _____

38. Define the following two terms.

    a. bradycardia _____

    b. tachycardia _____

39. What three factors determine pulse pressure?

    a. _____

    b. _____

    c. _____

40. Fill in the following blanks.

    a. A normal young adult's systolic blood pressure is _____ mmHg, or less.

    b. A normal young adult's diastolic blood pressure is _____ mmHg, or less.

    c. The systolic blood pressure of a baby may be as low as _____ mmHg.

    d. By middle age, normal blood pressure has risen to ____ over ____.

    e. During exercise, systolic blood pressure often rises as high as _____ mmHg.

____ 41. MAP is equal to all the following except

    a. average pressure driving blood through the arteries
    b. diastolic pressure plus one third of systolic pressure
    c. cardiac output times total peripheral resistance
    d. usually around 90 mmHg under normal conditions

42. Name the instrument that measures arterial blood pressure. _____

V. The Effects of Aging on the Blood Vessels (697)

1. The absence of what two lifestyle habits is conducive to many of circulatory system problems associated with age?

    a. _____

    b. _____

2. What are the two chief causes of cerebral hemorrhage?

   a. _____

   b. _____

VI. Developmental Anatomy of Major Blood Vessels (697)

1. Arrange the following parts of the embryonic heart and circulatory system in the proper order of development, beginning with a:

   a , ___, ___, ___, ___, ___, ___.

   a. atrium
   b. aortic arches
   c. atrioventricular orifice
   d. aortic sac
   e. truncus arteriosus
   f. bulbis cordis
   g. ventricle

2. How many pairs of aortic arches develop? _____

3. Which arch persists? _____

4. At what stage of development does the truncus arteriosus divide into the pulmonary trunk and the aorta? _____

VII. When Things Go Wrong (701)

1. What are five common sites for aneurisms?

   a. _____

   b. _____

   c. _____

   d. _____

   e. _____

2. What are three ways in which atherosclerosis can cause a heart attack?

   a. _____

   b. _____

   c. _____

3. What is usually the underlying cause of coronary artery disease?_____

___ 4. Which of the following has a positive correlation with the incidence of CAD?

   a. the ratio of chylomicrons to triglycerides
   b. the ratio of VLDL to total cholesterol
   c. the ratio of total cholesterol to LDL
   d. the ratio of HDL to LDL
   e. the ratio of LDL to total cholesterol

___ 5. Which of the following has a negative correlation with the incidence of CAD?

   a. the ratio of LDL to HDL
   b. the ratio of cholesterol in the diet to carbohydrates
   c. the ratio of HDL to total cholesterol
   d. the ratio of omega-3 oils to VLDL
   e. none of the above

6. What are two types of hypertension?

   a. _____

   b. _____

7. What are five ways a CVA can occur?

   a. _____

   b. _____

   c. _____

   d. _____

   e. _____

8. Describe the relationship between the following terms: thrombophlebitis, thrombus, embolus. _____

9. What results when the ductus arteriosus fails to shut at birth? _____

_____

Key Terms

aorta 654
arteriole 655
artery 654
baroreceptors 689
blood flow 684
blood pressure 684
blood velocity 685
capillary 654, 655
capillary fluid shift 691
cardioregulatory center 689
cerebral arterial circle 663
diastolic pressure 694

ductus arteriosus 698
hepatic portal system 663
hydrostatic pressure 688, 691
hypothalamic-hypophyseal portal
    system 663
mean arterial pressure 685
microcirculation 657
pulmonary circulation 660
pulmonary trunk 654
pulse 693
pulse pressure 694
resistance to blood flow 685, 692

systemic circulation 660
systolic pressure 694
total peripheral resistance 684
tunica adventitia 655
tunica intima 654
tunica media 654
vasomotor center 689
vein 654, 658
venous pump 692
venule 658

Post Test

Matching

____ 1. Some have relatively thick, elastic walls

____ 2. Drain capillary beds

____ 3. Distribute blood to various organs and body parts

____ 4. Have precapillary sphincters

____ 5. Conduct oxygenated blood to the heart

____ 6. Conduct deoxygenated blood to the heart

____ 7. The smallest category of muscular blood vessels

a. arteries
b. veins
c. arterioles
d. venules
e. metarterioles

____ 8. In arteries, composed mainly of collagenous and elastic fibers

____ 9. The central canal of any tubular structure

____ 10. The endothelial lining of an artery or a vein

____ 11. The thickest layer of large arteries

____ 12. "Vessels of vessels"

a. lumen
b. tunica media
c. tunica intima
d. tunica adventitia
e. vasa vasorum

Letting red stand for oxygenated blood, blue for deoxygenated blood, and C for capillary bed, match the sequences shown below.

____ 13. Heart→blue→C→red→heart

____ 14. Heart→red→C→blue→heart

____ 15.

Heart→red→C→blue→C→blue→heart

____ 16. Heart→red→heart→C→blue→heart

____ 17. Heart→red→red/blue→heart

a. coronary circuit
b. systemic circuit
c. pulmonary circuit
d. portal system
e. AV anastomosis

Matching

____ 18. Gives off branches to liver and spleen

____ 19. Distributes blood to arteries supplying cerebrum

____ 20. Formed by merging of left and right vertebral arteries

____ 21. Gives off branches to kidneys and adrenal glands

____ 22. Leads into a femoral artery

____ 23. Supplies one of the lower limbs

____ 24. Supplies the eyes and part of the cerebrum

a. basilar artery
b. internal carotid artery
c. common iliac artery
d. descending aorta
e. cerebral arterial circle

Identify the blood vessels indicated in the figure below.

___ 25. common carotid
___ 26. renal artery
___ 27. femoral artery
___ 28. brachiocephalic artery
___ 29. ulnar artery
___ 30. thoracic artery
___ 31. brachial artery
___ 32. subclavian artery
___ 33. common iliac artery

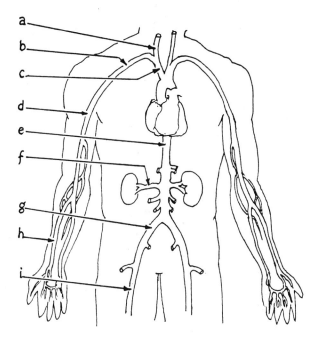

Identify the vessels in the figure below.

___ 34. hepatic portal vein
___ 35. great saphenous vein
___ 36. superior vena cava
___ 37. hepatic vein
___ 38. external jugular vein

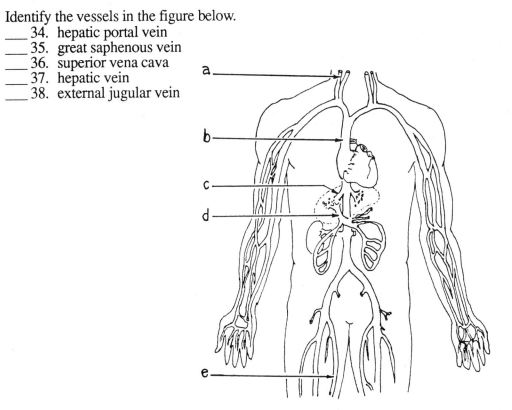

Multiple Choice

____ 39. The diaphragm receives blood from the
    a. posterior intercostal arteries
    b. celiac artery
    c. superior and inferior phrenic arteries
    d. superior mesenteric artery
    e. bronchial arteries

____ 40. The heart wall receives blood from the
    a. right brachial artery
    b. left and right coronary arteries
    c. superior vena cava
    d. left and right atria
    e. left and right ventricles

____ 41. The pancreas and most of the small intestines are supplied with blood by the
    a. esophageal and superior phrenic arteries
    b. common iliac arteries
    c. inferior mesenteric artery
    d. celiac and superior mesenteric arteries
    e. thoracic and lumbar arteries

____ 42. The superior and lateral thoracic arteries and the subscapular artery are branches of the
    a. axillary artery
    b. brachial artery
    c. radial artery
    d. circumflex humeral artery
    e. none of the above

____ 43. The veins of the face, neck, and thyroid gland are drained by the
    a. external jugular veins
    b. internal jugular veins
    c. common carotid arteries
    d. innominate arteries
    e. temporal veins

____ 44. Because arteries near the heart are elastic, blood flow through them is
    a. maintained at a uniform pressure
    b. increased
    c. decreased
    d. pulsatile
    e. slowed down

____ 45. Blood pressure is computed as the product of
    a. blood velocity times cardiac output
    b. $BP \times R$
    c. blood flow times resistance to flow ($F \times R$)
    d. $\Delta P \times R$
    e. $\Delta P \times CO$

____ 46. Baroreceptors in the the aorta and carotid sinus stimulate
    a. the cardiac center in the medulla oblongata
    b. a decrease in parasympathetic stimulation of the heart
    c. an increase in sympathetic stimulation of the heart
    d. an increase in arterial pressure
    e. both c and d

___ 47. The regulation of blood flow in circumscribed capillary beds in response to local changes in cellular metabolism is conceived of as

    a. hormonal regulation
    b. nervous regulation
    c. a baroreceptor response
    d. vasomotion
    e. autoregulation

___ 48. Hormones and/or enzymes that play an important role in the regulation of blood pressure include

    a. renin and angiotensin I
    b. angiotensins I and II
    c. vasopressin and bradykinin
    d. atriopeptin
    e. all of the above

___ 49. Chemoreceptors are stimulated by

    a. alterations in concentration of Ca^{2+}, Na^+, and K^+ in the blood plasma
    b. low concentration of O_2 and high concentrations of CO_2 and H^+
    c. changes in osmotic pressure (OP)
    d. vasopressin and bradykinin
    e. changes in hydrostatic pressure (HP)

___ 50. Which of the following mechanisms enable materials to be moved through capillary walls?

    a. diffusion
    b. filtration
    c. osmosis
    d. pinocytosis
    e. all of the above

51. List the five factors that have major significance in the return of venous blood to the heart.

a. _____

b. _____

c. _____

d. _____

e. _____

Integrative Thinking

1. Continuous capillaries, which lack the fenestrations of other types of capillaries, are the major kind of capillaries in skeletal muscle. How do you account for this?

2. When exposed to very cold weather, one's face flushes and, in fair-skinned persons, blushes with a warm glow. Depending upon how warmly dressed one happens to be, this flush is replaced in due course by an unpleasant sensation of coldness. One starts to shiver. A fair-complexioned person's face pales. Explain these responses to cold.

3. Mrs Theokilos is worried about her husband, Nicholas. He has frequent headaches and occasionally moments of dizziness. Nick's father died of cerebral hemorrhage at the age of 58, after several years of having similar symptoms. Nick is now 55 years old. Like his father, Nick is a heavy smoker. If you were Nick's physician, what would you look for first to explain his headaches and dizziness?

Your Turn

Choose either Option A or B.

Option A: This is an opportune time for small groups of young men and women to discuss objectively and dispassionately the problem of smoking. Appoint one of your group to serve as discussion leader, and another to take notes, in case the group is large and requires organization.

Option B: Let each member of your discussion group select a major organ (drawing from the old hat may be appropriate here) for a detailed anatomical description of its circulatory needs and features. The organs to choose from are the heart, lungs, kidneys, liver, and brain. The needs and features to consider are (a) acquisition of oxygen, fluids, nutrients, and any special substances; and (b) elimination of carbon dioxide, metabolic wastes, and any special cell products or metabolites.

22 The Lymphatic System

Active Reading

Introduction

1. What are two main differences between the circulatory and the lymphatic systems?

   a. _____

   b. _____

2. What are four major functions of the lymphatic system?

   a. _____

   b. _____

   c. _____

   d. _____

I. The Lymphatic System (709)

Match the following pairs.

___ 1. Tubular structures that converge to form drainage ducts   a. lymphatic capillaries
___ 2. Fluid bathing the interstitial compartment   b. lymphatic Vessels
___ 3. Closed-ended vessels   c. lymph nodes
___ 4. Open-ended vessels   d. tonsils
___ 5. Other organs of the lymphatic system   e. spleen and Thymus
___ 6. Lymphatic tissue of the throat region   f. lymph

7. At what point does the interstitial fluid take on the name lymph? _____

_____

8. List the factors that affect the flow of lymph.

    a. _____

    b. _____

    c. _____

    d. _____

Match the following pairs.

_____ 9. Upper right quadrant
_____ 10. Lower right quadrant
_____ 11. Upper left quadrant
_____ 12. Lower left quadrant

    a. thoracic duct
    b. right lymphatic duct
    c. lateral thoracic duct
    d. medial left lymphatic duct

13. How much lymph does the body contain? _____

14. What components of blood are absent from lymph? _____

15. What are the cells of lymph? _____

Match the following pairs.

_____ 16. Collective term for macrophages
_____ 17. Class of leukocytes capable of
developing into phagocytic cells
_____ 18. Tissue macrophages
_____ 19. Liver macrophages

    a. histocytes
    b. monocytes
    c. stellate reticuloendothelial cells
    d. reticuloendothelial system

20. What are the two fundamental types of lymphocytes?

    a. _____

    b. _____

21. Provide the functions for each of the following.

    a. lymphatic capillaries _____

    b. lymphatic vessels _____

    c. lymph nodes _____

    e. spleen _____

    f. thymus gland _____

    g. aggregate lymph nodules _____

22. What is the location of the cisterna chyli? _____

23. Define diapedesis. _____

24. How many cell layers thick are the walls of lymphatic capillaries? _____

25. What is the function of flap valves? _____

26. In what specialized lymphatic capillaries may chyle be found? _____

27. What are the two sets of lymphatics?

    a. _____

    b. _____

28. Where are the superficial lymph nodes located?

    a. _____

    b. _____

    c. _____

29. Where are the deep lymph nodes located?

    a. _____

    b. _____

    c. _____

    d. _____

    e. _____

30. On the figure below, locate the following: tonsils, cervical nodes, right lymphatic duct, supratrochlear nodes, thymus gland, axillary nodes, thoracic duct, spleen, appendix, inguinal nodes, lumbar nodes, iliac nodes.

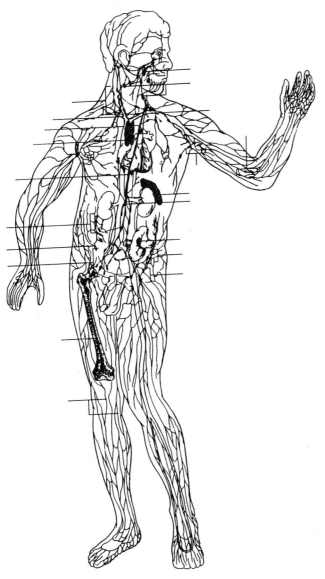

31. What areas of the body do the following nodes drain?

    a. supratrochlear _____

    b. intestinal _____

    c. lumbar _____

    d. iliac _____

    e. axillary _____

    f. cervical _____

    g. inguinal _____

32. Label all structures indicated in the figure below.

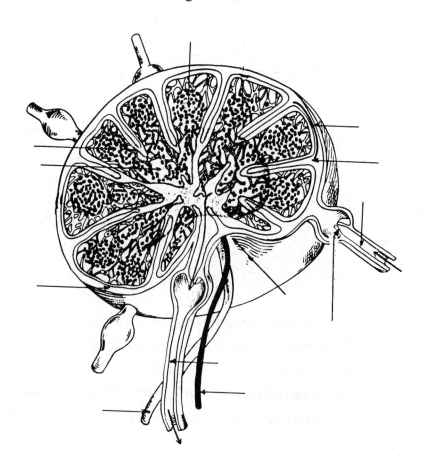

33. List and give the locations for the three types of tonsils.

    a. _____

    b. _____

    c. _____

34. What type of lymphatic vessels do the tonsils lack? _____

35. Why are the tonsils located where they are? _____

36. What is the largest lymphatic organ? _____

37. What are the two main functions of the spleen?

      a. _____

      b. _____

Match the following pairs.

  ___ 38. Remove damaged or dead erythrocytes   a. excess blood in spleen
  ___ 39. Released in response to loss of blood or bursts of   b. macrophages in spleen
           physical activity   c. red blood cells in spleen
  ___ 40. Activate lymphocytes for immune system   d. antigens in blood of spleen
  ___ 41. Produced during fetal life

42. What is the functional part of the spleen's medulla? _____

43. In what tissue are the venous sinusoids located? _____

44. How many lobes constitute the thymus gland? _____

45. When in life is the thymus gland most active? _____

46. What hormones does the thymus gland secrete? _____

47. In what two locations are Peyer's patches (aggregated lymph nodules) found?

      a. _____

      b. _____

48. Describe how the plasma cells generated in the aggregated lymph nodules are distributed

    along the length of the intestine. _____

49. What do the plasma cells produce? _____

50. Where do the GALT B cells become active? _____

II. Developmental Anatomy of the Lymphatic System (720)

1. On the figure below, identify the structures indicated.

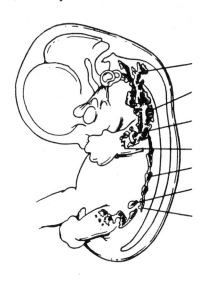

2. When does the development of the lymphatic system begin? _____

3. What fetal tissue develops into the thoracic duct? _____

4. What fetal tissue develops into spleen? _____

III. When Things Go Wrong (721)

1. The presence of what cells indicates Hodgkin's disease? _____

2. What are the first signs of Hodgkin's disease? _____

Key Terms

aggregated lymph nodules 715
aggregated unencapsulated lymph nodules 719
B cell 710
lymph 710
lymphatic capillaries 713
lymphatic system 710
lymphatic vessels 713
lymph node 713

lymphocyte 710
right lymphatic duct 713
spleen 716
T cell 710
thoracic duct 713
thymus gland 717
tonsils 715

Post Test

Matching

___ 1. A main channel for conveying lymph to the blood
___ 2. Characterizes the smallest lymphatic vessels
___ 3. Peyer's patches
___ 4. A final drainage destination for lymph
___ 5. The system for drainage of interstitial fluids

a. closed-ended
b. lymphatic vessels
c. thoracic duct
d. subclavian vein
e. aggregates of lymphoid nodules

___ 6. Specialized lymphatic capillaries in intestinal villi
___ 7. Mechanism of movement of white blood cells through capillary walls
___ 8. Structures located in the walls of lymphatic vessels and capillaries
___ 9. A homogeneous mixture of lymph and emulsified fat
___ 10. The dilated part of the thoracic duct at its origin

a. diapedesis
b. flap valves
c. lacteals
d. chyle
e. cisterna chyli

____ 11. The largest lymphoid organ

____ 12. Plays a major role in the early development of the immune system

____ 13. Located in the intestinal wall and appendix

____ 14. Similar to mucosa-associated lymphoid tissue (MALT)

____ 15. Widespread, strategically located, bean-shaped organs that filter the lymph

____ 16. Encapsulated aggregates of lymphoid nodules located in the general area of the throat

____ 17. A major site for the production of phagocytes and monocytes

____ 18. Serves as an important blood reservoir

____ 19. Site of conversion of undifferentiated lymphocytes to immunocompetent T cells

____ 20. Included are pharyngeal, palatine, and lingual types

a. lymph node
b. spleen
c. tonsils
d. Peyer's patches
e. thymus

Multiple Choice

____ 21. The vessel that drains lymph from the left side of the head and neck region

a. left lymphatic duct
b. right lymphatic duct
c. left jugular lymphatic duct
d. right jugular lymphatic duct
e. left jugular vein

____ 22. The vessel into which the vessel described in the preceding question drains directly

a. left common carotid artery
b. left jugular vein
c. thoracic duct
d. cisterna chyli
e. left subclavian vein

____ 23. The kind of cells that are produced in great numbers in the germinal centers of lymph nodes

a. lymphocytes
b. monocytes
c. macrophages
d. red blood cells
e. all of the above

____ 24. Lymph filtering through a lymph node exits

a. at various points along the capsule
b. by a single efferent vessel
c. by a series of afferent vessels
d. by an efferent vein
e. by medullary cords

____ 25. The adenoids are lymphoid aggregates located

a. at the point of entry of the auditory (Eustachian) tube
b. at the root of the tongue
c. on each side of the soft palate
d. in the upper posterior wall of the pharynx
e. on the medial side of the soft palate

Integrative Thinking

1. When a lymph node is overloaded, it hypertrophies (grows larger). What is the basis of the enlargement — edema?

2. If the axillary lymph nodes on the left side swell, what is indicated, and where would you look for the cause?

3. If you had a rather painful toothache, where would you look to see if it was due to an infection? If the toothache was accompanied by external swelling, what would this mean?

4. Why does a surgeon biopsy the lymph nodes adjacent to a neoplastic growth?

Your Turn

1. Time for discussion and making comparisons. Topic: (1) How the Lymphatic System Is Like the Blood Vascular System; (2) How the Lymphatic System Differs from the Blood Vascular System; and (3) How the Lymphatic System and Blood Vascular System Complement Each Other. Take notes, write it all up, and submit a report to your instructor, who might wish to make multiple copies to distribute to the entire class.

2. Do a similar analysis in which you compare the various lymphoid organs and tissues to each other. For example, what are the differences and similarities between the tonsils and the thymus, between the thymus and the spleen, between the spleen and lymph nodes, between lymph nodes and tonsils, etc.?

23 The Immune System

Active Reading

Introduction

1. What are pathogens? _____

2. What ability is at the core of the immune system? _____

3. What two broad classes of defenses does the body have to protect itself against invading pathogens?

   a. _____

   b. _____

I. Nonspecific Defenses of the Body (690)

1. What distinguishes the specific from the nonspecific defenses? _____

____ 2. Which of the following does <u>not</u> contribute to the skin's role in the immune system?

    a. high fat content
    b. ridges known as fingerprints
    c. acidic surface
    d. shedding of outer layer
    e. none of the above

____ 3. Which of the following is <u>not</u> a nonspecific defense?

    a. nasal hair
    b. sweat
    c. tears
    d. mucus
    e. none of the above

____ 4. What is the function of the cilia of the upper respiratory tract?

    a. sweep air along to aid in respiration
    b. sweep pathogens out of body
    c. sweep pathogens to the digestive tract
    d. sweep mucus to mouth and nose
    e. none of the above

___ ·5. Interferons are

    a. proteins
    b. lipids
    c. nonspecific immune cells
    d. the first of the specific defenses
    e. none of the above

Match the following pairs.

___ 6. Alpha interferon
___ 7. Beta interferon
___ 8. Gamma interferon

    a. made by fibroblasts
    b. made by T cells
    c. made by leukocytes

9. What distinguishes the cells that produce interferon? _____

10. How do interferons work? _____

_____

11. At what point do the classical pathway and the alternate pathway meet? _____

_____

12. What are the designations of the 20 or more inactive plasma proteins known as complement? _____

13. What is meant by "lysis"? _____

14. What do the following proteins help to promote?

    a. C1 _____

    b. C3a _____

    c. C3b _____

15. Describe the role of MAC in immune cytolysis. _____

_____

16. What are the localized effects of tissue damage and infection?

    a. _____

    b. _____

    c. _____

    d. _____

17. What causes each of the preceding?

    a. _____

    b. _____

    c. _____

    d. _____

18. What are pyrogens? _____

Match the following pairs.

___ 19. First stage of the inflammatory response
___ 20. Second stage
___ 21. Third stage
___ 22. Fourth stage
___ 23. Fifth stage
___ 24. Sixth stage
___ 25. Seventh stage
___ 26. Eighth stage
___ 27. Ninth stage
___ 28. Tenth stage
___ 29. Eleventh stage
___ 30. Twelfth stage

a. increased blood flow produces a characteristic redness and heat
b. after margination, the neutrophils pass through capillary walls and move toward microorganisms in the inflamed area
c. local inflammatory mediators are released
d. first stage of chemotaxis
e. repair of damaged tissue
f. each neutrophil may consume up to 20 bacteria before dying
g. opsonization
h. C3a, C5a, C5b, C6, and C7 form the membrane attack complex (MAT)
i. vasodilation
j. monocytes differentiate into macrophages
k. macrophages, monocytes, and histocytes begin to arrive on the scene
l. fibrinogen clots isolate healthy tissue

31. What is phagocytosis? _____

32. In the figure below, identify the following: leukocyte nucleus, lysosomes, microorganisms, phagocytic vesicle, phagolysosome, pseudopod.

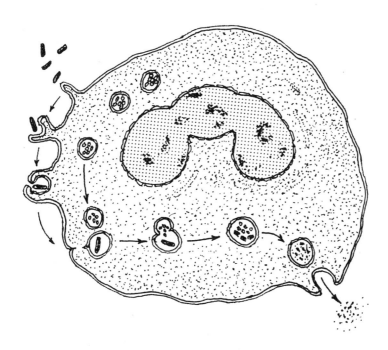

33. Where are defensins found? _____

II. Specific Defenses: The Immune Response (694)

1. How do the cells of the immune system "communicate" with each other? _____

   _____

2. What are the functional units of the immune system? _____

3. What are the two types of lymphocytes?

   a. _____

   b. _____

___ 4. An antigen is

   a. a foreign protein or polysaccharide molecule
   b. the functional unit of an antibody
   c. the precursor to a lymphocyte
   d. none of the above

___ 5. An epitope is

   a. an isotope used in immune system diagnosis
   b. the active site on an antibody
   c. the combining sites on an antigen
   d. none of the above

___ 6. Which of the following are capable of acting as an antigen?

   a. almost any kind of protein
   b. only proteins found on the surface of microorganisms
   c. only proteins and polysaccharides that are harmful to the body
   d. none of the above

___ 7. Antibodies are

   a. special lymphocytes that fight antigens
   b. B cells that fight antigens
   c. T cells that fight antigens
   d. none of the above

8. What part of an antibody is the constant region? _____

9. What part of an antibody is the variable region? _____

10. Which are the bivalent immunoglobulins?

    a. _____

    b. _____

    c. _____

11. What are the multivalent immunoglobulins?

    a. _____

    b. _____

Match the following pairs.

___ 12. The largest antibodies, these are a pentameter

___ 13. These are the most common type of immunoglobulin

___ 14. These are present only in small numbers

___ 15. These are found mainly in the mucous membrane of the nose and throat

___ 16. These are responsible for allergic reactions

a. IgA
b. IgD
c. IgE
d. IgG
e. IgM

17. Where in relation to antibodies do complement proteins bind to antigens? _____

18. What must be in place before the complement molecules will make a hole in the membrane of a foreign cell? _____

19. What are the central lymphoid tissues? _____

20. Where do newly formed, inactive lymphocytes migrate to? _____

21. Which type of lymphocyte secretes antibodies? _____

22. Complete the following table.

	T Cells	B Cells
Site of production of undifferentiated cell		
Site of differentiation		
Response after binding to antigen		
Antibody production		
Type of immunity produced		
Cytotoxic activity		
Factor causing response to antigens		
Effect on macrophages		
Basic functions		

23. What do B cells do when encountering a specific antigen for the first time?_____

_____

24. Against what type of antigens is antibody-mediated (humoral) immunity most effective?

_____

25. Why is the secondary immune response so much faster than the primary immune response? _____

26. What are the four ways by which T cells protect?

    a. _____

    b. _____

    c. _____

    d. _____

27. Why is the response of T cells called cell-mediated immunity? _____

28. Against what types of antigens does cell-mediated immunity work? _____

29. What are the five components of the clonal selection theory of antibody production?

    a. _____

    b. _____

    c. _____

    d. _____

    e. _____

30. What determines whether a cell is identifiable as self or as nonself? _____

31. Fill in the following table.

Type of T Cell	Function

III. Hypersensitivity (Allergy) (704)

1. What causes allergies? _____

2. In the case of allergies, what is the causative antigen known as? _____

3. What is degranulation ? _____

4. What are effects of the following?

    a. histamine _____

    b. leukotriene _____

5. What type of allergen entry into the body produces systemic reactions? _____

6. What type of cells are responsible for delayed hypersensitivity? _____

7. What type of tissue can easily be transplanted without being rejected? _____

8. What is the acceptance of foreign tissue as self called? _____

9. Why is it important to administer immunosuppressive drugs at the time of tissue transplantation? _____

IV. Autoimmunity and Immunodeficiency (707)

1. What is the basic cause of autoimmunity? _____

2. Complete the following table.

Disease	Specificity of Autoantibodies Against	Result
	Platelets	
	T cells	
	Macrophages, T cells	
	Adrenal gland	
	DNA, nuclear antigens	
	Thyroid antigens	
	Erythrocyte antigens	
	Immunoglobulin	
	Basement membrane	
	Intrinsic factor	
	Acetylcholine receptors on skeletal muscle	

3. What are some conditions that decrease the effectiveness of the immune system? _____

   _____

4. What is an example of an acquired immunodeficiency disease? _____

5. What is an example of an inherited immunodeficiency disease? _____

V. Types of Acquired Immunity (709)

1. What is innate resistance? _____

2. What is acquired immunity? _____

3. What is an example of natural passive immunity? _____

4. What is an example of artificial immunity? _____

5. What is active immunity? _____

6. When is artificial passive immunity used? _____

7. What kind of immunity does the injection of a vaccine promote?

VI. Monoclonal Antibodies (711)

1. What does a hybridoma cell produce? _____

2. What are monoclonal antibodies currently being used for? _____

VII. When Things Go Wrong (712)

1. What cells does HIV attack? _____

2. Which of the two forms of HIV is more severe and accounts for most AIDS cases in the United States? _____

3. What two diseases signaled the existence of AIDS in 1981?

   a. _____

   b. _____

4. Why is the AIDS virus called a retrovirus? _____

5. Where is the viral DNA of the AIDS virus assembled? _____

6. What virus causes infectious mononucleosis? _____

Key Terms

acquired immunity deficiency syndrome 747, 751
acquired immunity 747
active immunity 748
allergy 743
antibody 734
antibody-mediated immunity 737
antigen 730
antigen-antibody complex 730
autoimmunity 746
cell-mediated immunity 737
clonal selection theory 738

complement 725
cytotoxic T cell 742
delayed hypersensitivity T cells 742, 745
helper T cells 741
human immunodeficiency virus (HIV) 751
hypersensitivity 743
immune response 730
immunodeficiencies 747
immunoglobulin 734
inflammatory response 727, 737

innate resistance 747
interferons 725
memory cell 737
monoclonal antibody 749
nonspecific defenses 725
passive immunity 748
phagocytosis 729
plasma cell 735, 737
specific defenses 725, 730
suppressor T cell 741
T4 cells 751

Post Test

Multiple Choice

___ 1. Any organism that can cause disease is termed a

    a. virus
    b. bacterium
    c. fungus
    d. pathogen
    e. microorganism

___ 2. What kind of defense against infection are skin and mucous membranes and antiviral and anti-bacterial substances, such as the interferons?

    a. specific
    b. nonspecific
    c. chemical
    d. physical
    e. incidental

___ 3. Sweat, tears, lysozymes, and gastric acidity are considered to function as _____ barriers against invading microorganisms.

    a. chemical
    b. physical
    c. antiviral
    d. antibacterial
    e. inflammatory

___ 4. Small protein molecules produced by virus-infected cells defines

    a. interferon
    b. complement
    c. antibodies
    d. MAC
    e. pyrogens

___ 5. A group of inactive proteins pre-formed in blood plasma that enhance the body's defense systems

    a. interferon
    b. complement
    c. antibodies
    d. MAC
    e. pyrogens

Matching

___ 6. A group of related antibiotic proteins located inside neutrophils

___ 7. Protein molecules produced by fibroblasts

___ 8. They raise body temperature by their effect on the temperature regulatory center in the hypothalamus

___ 9. Produced by leukocytes, they bind to surface receptors of uninfected cells adjacent to infected ones

___ 10. Made by T cells

    a. alpha interferons
    b. beta interferons
    c. gamma interferons
    d. pyrogens
    e. defensins

___ 11. Process of initiating the proliferation of specific T cells and/or specific antibodies	a. opsonization
	b. membrane attack complex (MAC)
	c. cytolysis
___ 12. Cell destruction	d. antigen immunogenicity
___ 13. Immune adherence, promoting phagocytosis	e. antigenic determinants
___ 14. Specific combining sites on antigen surface	
___ 15. Produces holes in the cell membranes of invading microorganisms	

Using the list that follows, identify the various structures indicated on the diagram below.

___ 16. heavy chain, constant region
___ 17. disulfide bonds
___ 18. carbohydrate group
___ 19. hinge region
___ 20. light chain, variable region
___ 21. antigen-binding site
___ 22. light chain, constant region
___ 23. heavy chain, variable region
___ 24. binding site for complement
___ 25. mast cell attachment site

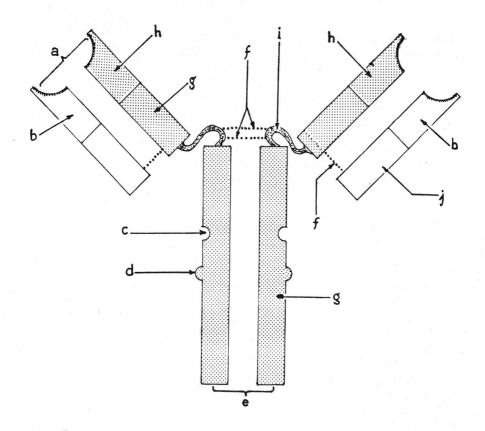

Matching

___ 26. Most common and abundant type of immunoglobulin
___ 27. Present only in small numbers, their specific function is not known
___ 28. The largest antibody
___ 29. The only antibody that passes through the placental barrier
___ 30. Composed of five 4-chain units and ten antigen-binding sites
___ 31. Responsible for immediate allergic reactions
___ 32. Major antibody class in saliva, tears, and human milk
___ 33. Phagocytes that become coated with this antibody demonstrate increased phagocytic efficiency
___ 34. Trigger the release of histamine from mast cells
___ 35. Synthesized during fetal life, they are the major antigen fighters

a. IgA molecules
b. IgD molecules
c. IgE molecules
d. IgG molecules
e. IgM molecules

Using the spaces provided below, arrange the following events in the order in which they occur during the development and resolution of a typical inflammatory response.

36. Answer: (__)→(__)→(__)→
    (__)→(__)→(__)

a. neutrophils, monocytes enter infected area
b. infecting organisms destroyed
c. capillary permeability increased, edema develops
d. microorganisms enter injured tissue area
e. damaged tissues repaired
f. vasodilation occurs, blood flow increases

Matching

___ 37. Attacks any invading microorganism; not antigen-specific
___ 38. Activates B cells by releasing B-cell growth factor
___ 39. Differentiate into plasma cells when exposed to antigen
___ 40. Ingest microorganisms
___ 41. Regulate activities of NK cells
___ 42. Secrete antibodies that mark foreign substances for destruction
___ 43. Release lymphokines
___ 44. Activate lymphokines
___ 45. Inhibit development of B cells into plasma cells
___ 46. The main cause of rejection of tissue or organ heterografts

a. B cells
b. delayed hypersensitivity T cells
c. helper T cells
d. suppressor T cells
e. natural killer (cytolytic, NK) T cells
f. plasma cells
g. macrophages

Multiple Choice

___ 47. Foreign substances that elicit high levels of IgE antibodies in the blood are known as

    a. allergens
    b. allergies
    c. suppressor T cells
    d. hypersensitives
    e. histamines

___ 48. The sequence of events that leads to the release of histamine during an allergic reaction is

    a. IgE coats allergen, allergen attacks mast cells, mast cells release histamine
    b. allergen stimulates suppressor T cells, plasma cells inhibited, IgE levels rise and coat mast cells, mast cells release histamine
    c. allergen stimulates plasma cells to produce IgE, IgE coats mast cells, allergen bonds to IgE, mast cells release histamine
    d. allergen stimulates IgE, IgE bonds to allergen, IgE-allergen complex attacks mast cells, mast cells release histamine

___ 49. In additions to histamine, mast cells and basophils release longer-lasting

    a. suppressor amines
    b. amine stimulants
    c. leukocytes
    d. leukotriene
    e. cytotoxins

___ 50. A severe allergic reaction may lead to life-threatening

    a. degranulation
    b. hypertension
    c. cardiac arrest
    d. anaphylactic shock
    e. both b and c

___ 51. Delayed hypersensitivity, as in the case of rejection of organ heterotransplants, is due mainly to the activities of

    a. B cells
    b. T cells
    c. plasma cells
    d. basophils
    e. eosinophils

Matching

___ 52. Results when the body produces its own antibodies or T cells in response to an antigen of foreign origin

    a. autoimmunity
    b. acquired immunodeficiency
    c. severe combined immunodeficiency
    d. acquired immunity
    e. natural active immunity
    f. passive immunity

___ 53. Antibodies produced against one's own tissues

___ 54. Immunity imparted by antibodies produced by another individual.

___ 55. Addison's disease

___ 56. Multiple sclerosis

___ 57. Pneumocystis carini pneumonia (PCP) and Karposi's sarcoma (KS) common sequellae

___ 58. X-linked failure to develop B or T cells during fetal life

___ 59. Produced by injection of a vaccine.

___ 60. AIDS.

Name four clinical tests for detection of HIV.

61. _____

62. _____

63. _____

64. _____

Integrative Thinking

1. A medical research company, working with scientists at Duke University, has identified genes that encode enzymes that inhibit certain human complement proteins. Using viral vectors, researchers have inserted these genes into the livers of pigs, which they hope (the livers, not the pigs) can be transplanted into human patients whose own livers have been destroyed by disease. Explain the rationale of these experiments.

2. A patient was identified as having increased titers (levels) of monoclonal IgM. What does this finding imply?

3. Raoul Benedict complains of itchy eyes, a sneezy, drippy nose, and chest wheezes. (a) What ails Raoul, and (b) what immunoglobulin would you predict to be involved?

Your Turn

Describe, step by step, all that happens from the time you cut your finger until the wound is healed.

24   The Respiratory System

Active Reading

Introduction

1. What are the components of the respiratory system?

   a. _____

   b. _____

   c. _____

Match the following pairs.

___ 2. Inspiration and expiration   a. cellular respiration

___ 3. Exchange of gases between blood   b. external respiration

       and lungs   c. internal respiration

___ 4. Utilization of oxygen as an electron   d. ventilation

       acceptor

___ 5. Exchange of gases in the body tissue

I.   General Functions of the Respiratory System (758)

1. What are the four steps of the respiratory system?

   a. _____

   b. _____

   c. _____

   d. _____

2. What is the main function of the respiratory system? _____

3. Name four additional functions of the respiratory system.

    a. _____

    b. _____

    c. _____

    d. _____

II. Anatomy of the Respiratory Tract (758)

1. What causes air to flow through the respiratory passages? _____

2. On the figure below, identify the following: nasal cavity, pharynx, larynx, bronchi, mediastinum, liver, trachea, diaphragm, bronchioles, heart.

3. What three actions are associated with air entering the nose?

    a. _____

    b. _____

    c. _____

4. List the function of each of the following structures.

    a. nasal cavity _____

    b. larynx _____

    c. diaphragm _____

    d. bronchioles _____

    e. paranasal sinuses _____

    f. trachea _____

    g. bronchi _____

    h. lungs _____

    i. pleurae _____

    j. pharynx _____

    k. alveoli _____

    l. goblet cells _____

    m. nasal epithelium _____

    n. olfactory epithelium _____

5. On the figure below, identify the following: superior concha, middle concha, inferior concha, frontal sinus, nasal cavity, sphenoidal sinus, internal naris, external naris, epiglottis, laryngopharynx, superior meatus, middle meatus, inferior meatus, soft palate, hard palate, tongue, mandible, hyoid bone, esophagus, oropharynx, uvula, thyroid cartilage, cricoid cartilage, trachea, palatine tonsil, auditory tube opening, fauces.

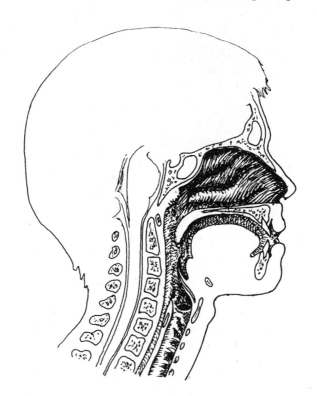

6. What is the proper order of the following, from superior to inferior? __, __, __

    a. oropharynx

    b. nasopharynx

    c. laryngopharynx

7. Describe the location of the three kinds of tonsils.

    a. _____

    b. _____

    c. _____

8. What anatomical structures support the larynx? _____

9. What holds the true vocal folds (cords) in place? _____

10. What factors control each of the following aspects of voice?

    a. pitch _____

    b. volume _____

    c. overtones _____

    d. individual variation _____

11. What is the function of the cilia in the trachea? _____

12. What are the five kinds of branches in the bronchial tree?

    a. _____

    b. _____

    c. _____

    d. _____

    e. _____

13. What makes up the conducting portion of the lung?

    a. _____

    b. _____

14. What makes up the respiratory portion of the lung?

    a. _____

    b. _____

    c. _____

    d. _____

15. On the figure below, identify the following: trachea, base, apex, inferior lobes, superior lobes, oblique fissures, horizontal fissure, bronchioles, primary bronchi, middle lobe, cardiac notch.

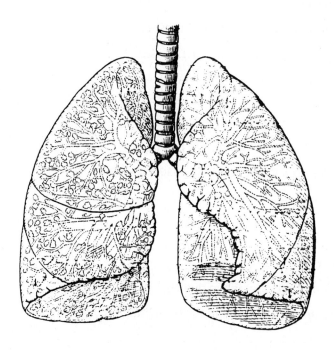

16. What is the functional unit of the lung? _____

17. What two types of cells make up the alveolar walls?

   a. _____

   b. _____

18. What supports the alveoli? _____

19. Describe the functions of the following.

   a. lung surfactant _____

   b. alveolar macrophages _____

20. How wide is each alveolus? _____

21. How thick is the alveolar wall? _____

22. Which lung is wider and shorter? _____

23. What make up the root of the lung?

   a. _____

   b. _____

   c. _____

____ 24. The left lung has
    a. one lobe
    b. two lobes
    c. three lobes
    d. four lobes

____ 25. The right lung has
    a. one lobe
    b. two lobes
    c. three lobes
    d. four lobes

____ 26. Which of the following is not a function of the pleura?
    a. to act as a lubricant for the lungs
    b. to provide innervation for the lungs
    c. to separate the lungs from other organs
    d. to aid in the mechanics of breathing

____ 27. What nerve carries parasympathetic innervation for constricting bronchial smooth muscle?
    a. bronchial nerve
    b. pulmonary nerve
    c. respiratory nerve
    d. vagus nerve
    e. none of the above

____ 28. Which circulation is involved in respiration?
    a. bronchial
    b. cardiac
    c. pulmonary
    d. cranial
    e. none of the above

____ 29. Which blood supply nourishes the lung tissue?
    a. bronchial
    b. cardiac
    c. pulmonary
    d. cranial
    e. none of the above

III. Physiology of Respiration (771)

1. What is the standard atmospheric pressure at sea level, (a) in mmHg _____, (b) in atmospheres _____?

2. What are three main chemical components of air and their percent concentration?

    a. _____

    b. _____

    c. _____

3. Describe Boyle's law. _____

_____

Match the following pairs.

___ 4. Pressure within the alveoli that changes with each breath

___ 5. Pressure within the rib cage that changes with each breath

___ 6. Pressure outside the epidermis that changes with altitude and approaching storm systems

a. atmospheric pressure
b. intraalveolar pressure
c. intrapleural pressure

7. At what times is the intraalveolar pressure always equal to atmospheric pressure? Why?

_____

___ 8. The intrapleural pressure

a. is always greater than atmospheric pressure
b. is always less than atmospheric pressure
c. is always equal to atmospheric pressure
d. fluctuates with respect to atmospheric pressure
e. none of the above

9. What is the major muscle used in quiet inspiration? _____

10. What muscles are used in quiet expiration? _____

11. What muscles cause forced expiration? _____

12. What is "compliance" in the context of respiration physiology? _____

13. Express compliance in terms of changes in lung volume and intraalveolar pressure.

_____

Match the following pairs.

___ 14. Expiratory reserve volume plus residual volume
___ 15. Inspiratory reserve volume plus tidal volume
___ 16. Vital capacity volume plus residual volume
___ 17. Inspiratory reserve volume plus tidal volume plus expiratory reserve volume

a. inspiratory capacity
b. vital capacity
c. Ffunctional residual capacity
d. total lung capacity

18. Describe the location and volume of the anatomical dead space. _____

19. What does MRV equal? _____

20. Why is it more efficient to breath deeply than to breath rapidly? _____

_____

21. Define each of the following terms.

    a. residual volume (RV) _____

    b. tidal volume (TV) _____

    c. IRV _____

    d. ERV _____

    e. IC _____

    f. FRC _____

    g. TLC _____

    h. VC _____

    i. ADS _____

    j. MRV _____

22. What is Dalton's law? _____

23. What is the partial pressure of oxygen in air at an atmospheric pressure of
700 mmHg? _____

24. True or false? The partial pressure of oxygen is greater in the alveoli than in the
blood. ____

25. True or false? The partial pressure of cardon dioxide is greater in the alveoli than in the
blood. ____

26. Describe Henry's law. _____

27. Which has a greater solubility coefficient, oxygen or carbon dioxide? _____

28. Why might deep sea divers breathe a mixture of helium and oxygen instead of nitrogen
and oxygen? _____

29. According to Fick's law, how would each of the following variables affect the rate of gas
diffusion across a membrane?

    a. low partial pressure differential _____

    b. high surface area for diffusion _____

    c. thin membrane _____

    d. low solubility of gas _____

    e. high molecular weight of gas _____

IV. Gas Transport (777)

1. What is external respiration? _____

2. What is internal respiration? _____

3. What is the partial pressure of oxygen in blood leaving the alveolar capillaries? _____

4. What percent of the blood's oxygen is carried by hemoglobin? _____

5. What is the most important factor in the loading and unloading of hemoglobin? _____

6. Describe how four other factors influence the loading and unloading of oxygen by hemoglobin.

    a. _____

    b. _____

    c. _____

    d. _____

7. What percent of the carbon dioxide in the blood is in the form of dissolved gas? _____

8. In what form is most of the carbon dioxide in the blood? _____

9. Describe the chloride shift. _____

10. What effect does the chloride shift have on the oxygen carrying capacity of the plasma?

    _____

11. Distinguish between the Haldane effect and the Bohr effect. _____

    _____

12. Give two reasons why carbon monoxide is dangerous.

    a. _____

    b. _____

V. Neurochemical Control of Breathing (782)

1. How are the rate and depth of breathing controlled? _____

2. By what secondary factors are the involuntary nerve impulses for respiration regulated?

    a. _____

    b. _____

    c. _____

2. What are the three subareas of the respiratory center, and where are they located?

    a. _____

    b. _____

    c. _____

3. What is the function of the DRG? _____

4. What is the function of the VRG? _____

5. Describe the functions of the following.

    a. pneumotaxic center _____

    b. apneustic center _____

6. What factors cause the following.

    a. coughing or sneezing _____

    b. holding one's breath _____

    c. Hering-Breuer reflex _____

7. Describe the process by which each of the following acts to increase ventilation.

    a. increased arterial hydrogen ion concentrations _____

    _____

    b. decreased arterial partial pressure of molecular oxygen _____

    _____

    c. increased arterial carbon dioxide partial pressure _____

    _____

VI. Other Activities of the Respiratory System (785)

1. What causes hiccups? _____

2. What is a cause of snoring? _____

VII. Effects of Aging on the Respiratory System (785)

1. What happens to alveolar tissue with age? _____

2. What happens to men's voices with age? _____

VIII. Developmental Anatomy of the Respiratory Tree (786)

1. Where does the bronchial bud form? _____

2. At what point in fetal life do the alveoli first appear? _____

3. At what age does the number of alveoli stabilize? _____

IX. When Things Go Wrong (787)

1. What is the most common form of rhinitis? _____

2. What is the primary symptom of an inflammation of the larynx? _____

3. What factors cause bronchial asthma? _____

4. What is the symptom of emphysema? _____

5. What is pneumonia and how is it caused? _____

6. What microorganism cause tuberculosis? _____

7. How do you account for the increasing incidence of tuberculosis? _____

    _____

8. Name the three stages or types of pleurisy.

    a. _____

    b. _____

    c. _____

9. What are the physiological effects of mountain sickness?

a. _____

b. _____

c. _____

d. _____

e. _____

f. _____

Key Terms

alveoli 766
bronchi 762, 764
bronchiole 764
cellular respiration 758
chloride shift 781
diaphragm 772
epiglottis 762
expiration 758, 771, 772
expiratory reserve volume 774
external intercostal muscles 772
external respiration 758, 777
functional residual capacity 774

hemoglobin 777
inspiration 758, 771, 772
inspiratory capacity 774
inspiratory reserve volume 774
internal respiration 758
laryngopharynx 761
larynx 762
lung 766
medullary rhythmicity center 782
minute respiratory volume 774
nasal cavity 758
nasopharynx 761

oropharynx 761
pharynx 761
pleural cavity 768
residual volume 774
respiratory center 782
respiratory tract 758
tidal volume 774
total lung capacity 774
trachea 762
ventilation 758
vital capacity 774

Post Test

Matching

____ 1. "The sum of biochemical events by which the chemical energy of food is released to provide energy for life's processes"

____ 2. The expandibility of the thoracic cage

____ 3. The exchange of gases in the body tissues

____ 4. CO_2 from the cells of the body is exchanged for O_2 from the blood

____ 5. The exchange of gases between the lungs and the blood

____ 6. The mechanical process that moves air into and out of the lungs

____ 7. Oxygen is used as a final electron acceptor in the mitochondria

____ 8. Includes inspiration and expiration

____ 9. The change in lung volume per unit change in intraalveolar pressure

____ 10. Oxygen in the alveoli diffuses into the blood, and carbon dioxide diffuses from the blood into the alveoli

a. cellular respiration
b. external respiration
c. internal respiration
d. ventilation
e. compliance

Multiple Choice

____ 11. The nasal chamber is where

a. inhaled air is warmed and moistened
b. the olfactory receptors are situated.
c. vocal sounds are amplified and modulated
d. inhaled air is filtered
e. all of the above

____ 12. The lateral walls of the nasal chamber form the

a. superior, middle, and inferior meatuses
b. nasal conchae
c. cribriform plate
d. paranasal sinuses
e. olfactory epithelium

____ 13. The funnel-shaped passage, commonly called the throat, that connects the oral cavity to the respiratory passage and esophagus is technically known as the

a. nasopharynx
b. oropharynx
c. laryngopharynx
d. pharynx
e. larynx

____ 14. Situated in the nasopharynx are the

a. openings of the auditory (Eustachian)
  tubes
b. paired meatuses
c. fauces
d. palatine tonsils
e. false vocal folds

____ 15. The true vocal folds are located in
the walls of the

a. oropharynx
b. laryngopharynx
c. larynx
d. trachea
e. hyoid apparatus

Label the diagram below, which shows the internal aspect of the larynx.
____ 16. thyroid membrane
____ 17. cricoid cartilage
____ 18. true vocal fold
____ 19. vestibular vocal fold
____ 20. thyroid cartilage
____ 21. hyoid bone
____ 22. arytenoid cartilage
____ 23. epiglottis

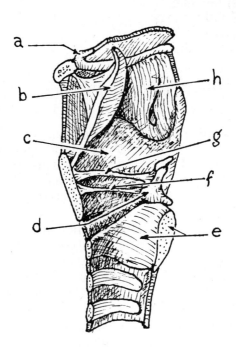

Label the diagram below.
___ 24. trachea
___ 25. thyroid cartilage
___ 26. secondary bronchus
___ 27. tertiary bronchus
___ 28. terminal bronchioles
___ 29. tracheal cartilages
___ 30. primary bronchus
___ 31. cardiac notch

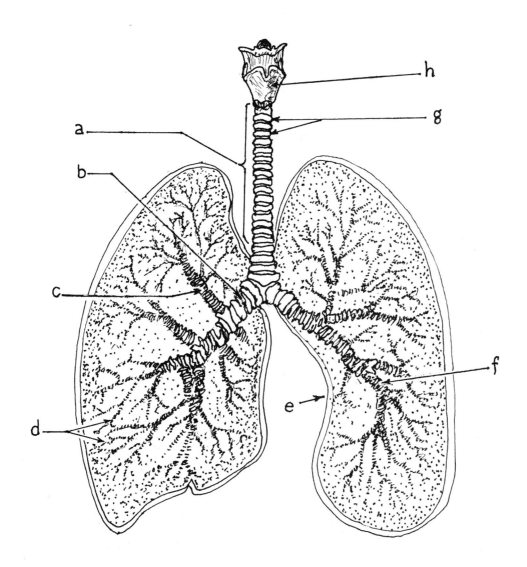

Matching

___ 32. The serous membrane that covers the lungs
___ 33. Subdivisions of lung lobes
___ 34. Area on the medial surface of each lung for entry of bronchi, blood vessels, and nerves
___ 35. Occupies midventral region of thoracic cavity
___ 36. Branches off from tertiary bronchi.
___ 37. Lines the thoracic cavity

a. hilus
b. medastinum
c. visceral pleura
d. parietal pleura
e. bronchopulmonary segments
f. bronchiole

Short Answer

38. From what circulation do the lungs receive oxygen-poor blood? _____

39. From what circulation do the lungs receive oxygen-rich blood? _____

Matching

___ 40. The volume of a gas varies inversely as its pressure.
___ 41. The law of gas diffusion, expressed as $V_{gas} = (P_1 - P_2) \times A/T \times D$.
___ 42. In a mixture of gases, the solubility of each in a liquid is proportional to its partial pressure and to its solubility coefficient.
___ 43. Each gas in a mixture of gases exerts pressure in proportion to its concentration in the mixture.

a. Dalton's law
b. Boyle's law
c. Henry's law
d. Fick's law

Complete the table below, using the appropriate letters: "a" for relaxed, "b" for weakly contracted, and "c" for strongly contracted.

	Action	External Intercostals	Diaphragm	Abdominal Muscles
44.	Quiet inspiration			
45.	Quiet expiration			
46.	Forced inspiration			
47.	Forced expiration			
48.	Holding a full breath			

Multiple Choice

___ 49. The maximum volume to which the lungs can be expanded is the

a. inspiratory capacity
b. tidal volume
c. functional residual capacity
d. vital capacity
e. total lung capacity

___ 50. The volume of air that is inhaled during quiet breathing while at rest is the

a. inspiratory capacity
b. tidal volume
c. vital capacity
d. residual volume
e. total lung capacity

___ 51. The partial pressure of a gas in a mixture of gases is calculated as the

a. percent of that gas in the mixture
b. concentration of that gas in the mixture
c. the percent concentration of that gas times the total pressure of the mixture
d. the total pressure of the mixture divided by the pressure of that gas
e. none of the above

___ 52. When oxygen becomes bound to hemoglobin, the compound that is formed is

a. methemoglobin
b. reduced hemoglobin
c. deoxyhemoglobin
d. oxyhemoglobin
e. hemoglobin dioxide

___ 53. When O_2 is released from hemoglobin, the hemoglobin molecule is designated as

a. HHb
b. $HbbO_2$
c. metaHb
d. $HbCl_2$
e. $HbCO_2$

___ 54. The most important factor influencing the binding and release of O_2 from hemoglobin is

a. Po_2
b. Pco_2
c. pH
d. temperature
e. 2,3-DPG

___ 55. Other factors that strongly favor the unloading of O_2 from hemoglobin are

a. increased pH
b. decreased pH
c. increased Pco_2
d. both a and c
e. both b and c

___ 56. Most of the CO_2 in the blood is carried in plasma as

a. molecular CO_2
b. CO_3^{2-}
c. HCO_3^-
d. carbonic acid
e. carbonic anhydrase

___ 57. The removal of bicarbonate ions from the erythrocytes results in a positive charge inside the cell, which is neutralized by the

a. enzyme carbonic anhydrase
b. chloride shift
c. dissociation of $NaHCO_3$ in the plasma
d. reaction $H^+ + HCO_3^- \rightarrow H_2CO_3$
e. $NaCl \rightarrow Na^+ + Cl^-$

Identify and locate on the diagram below the respiratory centers of the CNS.
___ 58. apneustic center
___ 59. medulla oblongata
___ 60. pons
___ 61. dorsal medullary rhythmicity center
___ 62. ventral medullary rhythmicity center
___ 63. pneumotaxic center

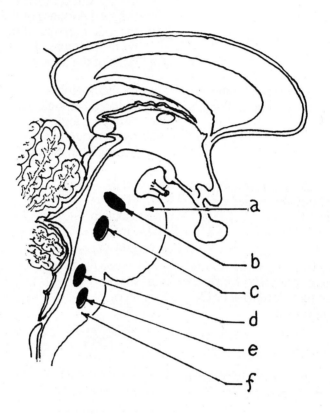

Multiple Choice

___ 64. The respiratory centers of the brainstem can be influenced by inputs from

a. centers in the hypothalamus
b. cortical areas, such as involved in speaking or swimming
c. chemoreceptors affected by air pollutants
d. inflation reflexes, such as the Hering-Breuer reflex
e. any or all of the above

___ 65. The medullary respiratory centers also receive input from

a. stimulated muscle and joint proprioceptors
b. elevated body temperature, such as a fever
c. changes in the partial pressure of O_2 in arterial blood
d. an increase in H^+ concentration in the blood
e. any or all of the above

Integrative Thinking

1. Shortness of breath is common among young and old alike. Outline all the causes of this complaint that you can think of.

2. The cabins of airplanes flying at high altitudes are pressurized. Why is this?

3. Scuba divers used to fill their tanks with a mixture of oxygen and helium, until they learned that helium is often contaminated with low levels of radioactivity. Why did they do this?

Your Turn

You and the rest of your study group are ready by this time to try what is called "Concept Mapping." The respiratory system is great for this. It would help if you could meet in a room with a large blackboard, but if none is available, each of you should have an 81/2 x 11 plain paper pad. Here is the procedure. Let's use a car as an example: In the middle of your writing surface write the name of some part of the car — any part. Let's say the radio. Put a circle around it. What is the radio attached to? The dashboard. Put it above or below the radio and connect them with a line. What is the dashboard attached to? The firewall, perhaps, or the windshield. Put one of them down and connect it to the dashboard. Better yet, put them both down and make appropriate connections. Now you're on your way. By the time you get to the spark plugs or the wheel covers you should have a pretty good idea of what a car is. O.K. Start! Did somebody suggest trachea? Or was it pneumotaxic center? My hearing is not so good these days. Better write it down. And put a circle around it! Next time we'll include concepts of function; for now, just structure will do.

25 The Digestive System

Active Reading

Introduction

Match the following pairs.

___ 1. The body converts large nutrient molecules into small molecules.
___ 2. The body eliminates solid waste.
___ 3. The body takes in food through the mouth.
___ 4. Usable nutrient molecules pass from the intestines into the bloodstream and lymphatic system.
___ 5. Involuntary smooth muscle contractions move material through the digestive tract.

a. ingestion
b. peristalsis
c. digestion
d. absorption
e. defecation

I. Introduction to the Digestive System (796)

1. What are the organs of the alimentary canal?

a. _____

b. _____

c. _____

d. _____

e. _____

f. _____

g. _____

h. _____

i. _____

2. Which of the organs named above make up the gastrointestinal (GI) tract?

     a. _____

     b. _____

     c. _____

     d. _____

     e. _____

     f. _____

3. What organs make up the associated structures of the digestive system?

     a. _____

     b. _____

     c. _____

     d. _____

     e. _____

     f. _____

     g. _____

     h. _____

4. What are the four tissue layers of the digestive system?

     a. _____

     b. _____

     c. _____

     d. _____

II.   Mouth (796)

1. When you inflate your cheeks, what cavity are you expanding?_____

2. In the figure below, identify the following: superior red free margin, inferior red free margin, superior labial frenulum, inferior labial frenulum, lingual frenulum, tongue, palatine tonsils, palatopharyngeal arch, palatoglossal arch, hard palate, soft palate, uvula, fauces, gingiva.

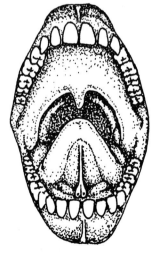

3. Describe the location of the oral cavity proper. _____

___ 4. The gag reflex follows stimulation of the

   a. uvula
   b. vestibule
   c. deciduous teeth
   d. fauces
   e. all of the above

___ 5. Which of the following functions does saliva serve?

   a. converts starches into sugar
   b. intensifies taste of food
   c. aids in movement of food through esophagus
   d. begins process of chemical digestion
   e. all of the above

___ 6. The skin of the red free margin contains

   a. translucent epidermis
   b. sebaceous glands
   c. sweat glands
   d. hair follicles
   e. all of the above

___ 7. A thick, stratified squamous epithelium lines the

   a. deciduous teeth
   b. red free margin
   c. labial frenulum
   d. cheeks
   e. all of the above

___ 8. How many deciduous teeth will a normal child have?

   a. 4
   b. 6
   c. 10
   d. 20
   e. none of the above

___ 9. How many permanent teeth will a normal adult have?

   a. 16
   b. 20
   c. 32
   d. 40
   e. none of the above

10. What are the four kinds of permanent teeth?

   a. _____

   b. _____

   c. _____

   d. _____

11. What holds teeth in their sockets? _____

12. On the figure below, identify the following: cellular cement, crown, dentine, periodontal ligament, root canal, apical foramen, neck, root, gingiva, enamel, bone of jaw.

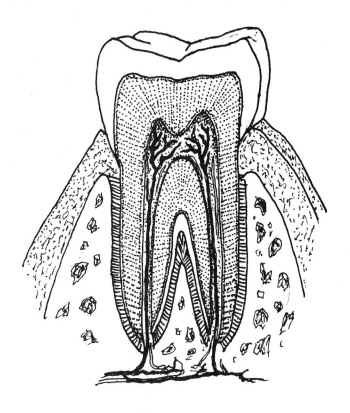

13. True or false? The tongue aids in digestion. ___

14. What are the two sections of the tongue's mucous membrane?

    a. _____

    b. _____

15. What nerve innervates the tongue? _____

16. What are the three types of papillae on the tongue? Where is each type located?

    a. _____

    b. _____

    c. _____

17. What are the two sections of the palate?

    a. _____

    b. _____

18. What prevents food from entering the nasal cavity during swallowing? _____

19. What are the three largest pairs of salivary glands?

    a. _____

    b. _____

    c. _____

20. What are some of the functions of saliva? _____

21. Why is the secretion of the parotids glands more clearer and more watery than other saliva? _____

22. Describe the location of the submandibular ducts. _____

23. Of the three major pairs of salivary glands, which secretes saliva highest in mucin and viscosity? _____

____ 24. Which of the following is absent in saliva?

    a. water
    b. salivary amylase
    c. phospholipids
    d. electrolytes
    e. mucin

____ 25. Which of the following does saliva <u>not</u> assist?

    a. digestion of proteins and fat
    b. oral hygeine
    c. taste
    d. speech
    e. digestion of carbohydrates

____ 26. Which of the following does not induce the production of saliva?

    a. chewing
    b. smell
    c. physical exertion
    d. mental images of chocolate
    e. the sight of chocolate

____ 27. Which of the following results in a thick, mucin-rich saliva?

    a. efferent ouput via the parasympathetic fibers
    b. activation of the sympathetic division of the autonomic nervous system
    c. afferent input from the pressoreceptors in the mouth
    d. afferent input from the chemoreceptors in the mouth
    e. none of the above

III. Pharynx and Esophagus (806)

1. Where does the voluntary phase of swallowing take place? _____

2. What are the two functions of the pharynx?

    a. _____

    b. _____

3. On the figure below, identify the following: nasopharynx, oropharynx, laryngopharynx, esophagus, trachea, epiglottis, soft palate.

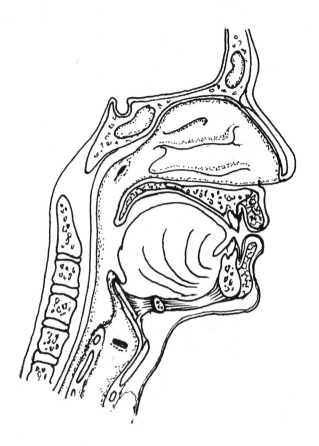

4. How long is the esophagus? _____

5. What closes the superior esophageal sphincter? _____

6. What produces heartburn? _____

8. Define the following terms.

    a. bolus _____

    b. deglutition _____

    c. peristalsis _____

9. Complete the following table.

Stage of Swallowing	Location of Bolus	Muscles Involved
Voluntary oral phase		
Involuntary pharyngeal phase		
Involuntary esophageal phase		

10. What are borborygmus and eructation? _____

IV. Abdominal Cavity and Peritoneum (809)

1. Describe the locations of the following structures.
    a. parietal peritoneum _____
    b. visceral peritoneum _____
    c. peritoneal cavity _____
    d. serous fluid _____
    e. retroperitoneal abdominal organs _____

2. What is the function of mesenteries? _____

3. Where is the fat in a "pot belly" located? _____

4. What is the function of the greater omentum? _____

5. Where is the lesser omentum located? _____

V. Stomach (810)

1. How is the location of the stomach described? _____

2. On the figure of the stomach below, identify the lesser curvature, greater curvature, esophagus, cardiac orifice, fundus, cardiac region, pyloric sphincter, body, rugae, pyloric region, pyloric canal, duodenum.

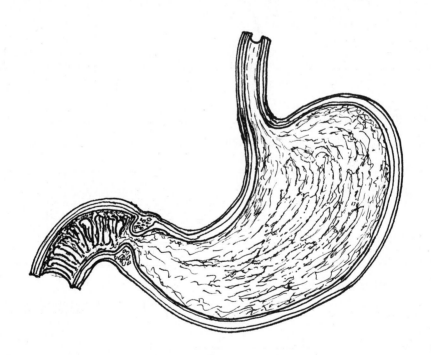

3. What are the three muscle layers of the stomach?

     a. _____

     b. _____

     c. _____

4. What enables the stomach to expand and contract so much? _____

5. How many gastric glands extend from each gastric pit? _____

6. What are the three functions of the stomach?

     a. _____

     b. _____

     c. _____

Match the following pairs.

____   7. First type of mixing, a few minutes   a. strong peristaltic waves
         after food enters stomach   b. pyloric pump
____   8. Second type of mixing, once pyloric   c. slow peristaltic mixing waves
         region begins to fill
____   9. Mechanism that forces chyme past
         the pyloric sphincter

10. List one physical condition, two emotional conditions, and three chemical conditions that increase the rate of gastric emptying.

     a. _____

     b. _____

     c. _____

     d. _____

     e. _____

     f. _____

11. List three emotional conditions, three chemical conditions, and two physical conditions that inhibit the rate of gastric emptying.

     a. _____

     b. _____

     c. _____

     d. _____

     e. _____

     f. _____

     g. _____

     h. _____

12. How is pepsin produced? _____

13. What protects the stomach wall from being digested? _____

14. Complete the following table.

	Stimulus(i)	*Secretion(s)*
Cephalic phase	Food is seen, smelled, tasted, chewed, or swallowed	
Gastric phase		Increased gastric secretion
Intestinal phase, excitatory component		
Intestinal phase, inhibitory component		

VI. Small Intestine (817)

1. Where does alcohol become absorbed into the bloodstream? _____

2. How long does food stay in the small intestine? _____

3. What are the three parts of the small intestine?

   a. _____

   b. _____

   c. _____

4. On the figure below, identify the following: esophagus, stomach, duodenum, jejunum, ileum, ileocecal valve, ascending colon, transverse colon, descending colon.

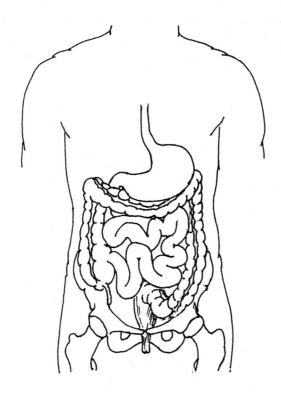

5. How does each of the following aid in the absorption of nutrients?

    a. plicae circulares _____

    b. villi _____

    c. microvilli _____

6. What is the total surface area of the small intestine? _____

7. Complete the following table of intestinal cell types.

Cell Type	Location	Function
Columnar absorptive cells		
Undifferentiated cells		
Mucous goblet cells		
Paneth cells		
Enteroendocrine cells		

8. What is the function of each of the following phenomena?

    a. segmenting contractions _____

    b. peristaltic contractions _____

9. What system of the body controls and regulates the small intestinal contractions? _____

10. What is succus entericus? _____

11. Describe the function of the three classes of digestive enzymes of the small intestine.

    a. _____

    b. _____

    c. _____

12. What are the nine steps of lipid absorption?

    a. _____

    b. _____

    c. _____

    d. _____

    e. _____

    f. _____

    g. _____

    h. _____

    i. _____

13. Complete the following table.

Substance	Site(s) of Absorption	Mechanism(s) of Absorption
Carbohydrates		
Proteins		
Water		
Fat-soluble vitamins		
Water-soluble vitamins		
Nucleic acids		

14. Which of the major ions and trace elements utilize active transport for absorption in the small intestine?

a. _____

b. _____

c. _____

d. _____

e. _____

VII. Pancreas as a Digestive Organ (825)

1. In the figure below, identify the stomach, liver, hepatic ducts, duodenum, jejunum, pancreas head, pancreas body, pancreas tail, kidneys, spleen, main pancreatic duct, secondary pancreatic duct.

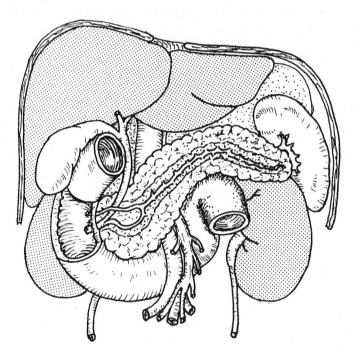

2. How are the exocrine cells grouped? _____

3. Describe the functions of the following enzymes.

   a. pancreatic lipase _____

   b. pancreatic amylase _____

   c. pancreatic proteolytic enzymes _____

4. Complete the following table.

Secretion	Source(s)	Place(s) of Digestion	Enzyme(s)	pH	Digestive Function
Saliva					
Mucus					
Gastric juice					
HCl					
Bile					
Intestinal juice					
Pancreatic juice					

5. What is the effect of the following hormones on the pancreas?

   a. secretin _____

   b. CCK _____

VIII.   Liver as a Digestive Organ (827)

1. In the figure below, identify the following structures: left lobe of liver, right lobe of liver, falciform ligament, ligamentum teres, gallbladder, diaphragm.

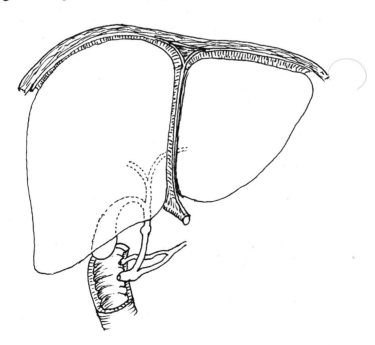

Match the following pairs.

____ 2. Supplies liver with oxygenated
blood, about 20% of the liver's
blood supply

____ 3. Drains blood from the liver into the
inferior vena cava

____ 4. Supplies blood to liver from the
entire GI tract, about 80% of the
liver's blood supply

a. hepatic artery
b. hepatic portal vein
c. hepatic vein

5. Describe the path of bile from the liver cells to the second part of the duodenum. _____

____ 6. Liver cells are known as

a. lobules
b. sinusoids
c. hepatic cells
d. portal cells
e. none of the above

____ 7. The functional units of the liver are

a. lobules
b. sinusoids
c. hepatic cells
d. portal veins
e. none of the above

____ 8. The delicate blood channels in the
liver are

a. lobules
b. sinusoids
c. hepatic cells
d. portal veins
e. none of the above

____ 9. Lining the walls of the sinusoids are

a. lobules
b. smaller sinusoids (fractal geometry)
c. hepatic cells
d. portal veins
e. none of the above

10. Explain the role of the liver in

    a. carbohydrate metabolism _____

    b. lipid metabolism _____

    c. protein metabolism_____

    d. vitamin and mineral utilization _____

11. What is in bile?_____

IX. Gallbladder and Biliary System (832)

1. Describe the gallbladder's storage function. _____

2. What are the three layers of the gallbladder?

   a. _____

   b. _____

   c. _____

3. What are the four components of the biliary system?

   a. _____

   b. _____

   c. _____

   d. _____

4. What two sphincters are associated with the biliary system?

   a. _____

   b. _____

5. What triggers the release of bile from the gallbladder? _____

X. Large Intestine (833)

1. In the figure below, identify the following: ileum, ileocecal valve, cecum, ascending colon, right colic flexure, transverse colon, left colic flexure, descending colon, sigmoid colon, rectum, anus.

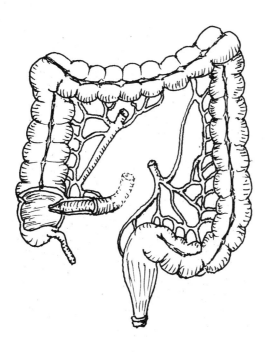

2. What are the three main structural differences between the large and small intestines?

    a. _____

    b. _____

    c. _____

3. What are the three main functions of the large intestine?

    a. _____

    b. _____

    c. _____

4. Why is the large intestine much less absorptive than the small intestine? _____

5. Which of the anal sphincters is under voluntary control? _____

6. What positive role do the bacteria of the large intestine fulfill? _____

7. What is the ratio of water to solids in feces? _____

8. What is responsible for the foul smell of feces? _____

9. What accounts for the movement of materials through the large intestine? _____

10. Describe the process of defecation.

    a. _____

    b. _____

    c. _____

    d. _____

    e. _____

    f. _____

XI. The Effects of Aging on the Digestive System (839)

1. What is diverticulosis? _____

2. For what reasons are the aged often malnourished? _____

XII. Developmental Anatomy of the Digestive System (839)

1. What are the three specific regions of the primitive gut?

    a. _____

    b. _____

    c. _____

2. Into what does each of the regions develop?

    a. _____

    b. _____

    c. _____

XIII. When Things Go Wrong (840)

1. What result of anorexia may prove fatal? _____

2. What is the effect of the decreased cholecystokinin secretion that may occur in bulimarectics? _____

3. What causes cholelithiasis? _____

4. What disease of the liver is prevalent among alcoholics over the age of 50? _____

5. What is a dietary cause of colon-rectal cancer, the second most common form of cancer in the United States? _____

6. What is the gene known as adenomatous polyposis coli? _____

7. Does constipation increase the amount of toxins in the body? _____

8. What are some causes of diarrhea? _____

9. What are the symptoms of viral hepatitis? _____

10. What are the causes of nonviral hepatitis?

     a. _____

     b. _____

     c. _____

11. What is a hernia? _____

12. What part of the brain controls vomiting? _____

13. What are some of the dangerous effects of excessive vomiting? _____

Key Terms

absorption 796, 821
acini 825
alimentary canal 796
anal canal 837
anus 837
bile 832
biliary system 832
bolus 807
cholecystokinin 827
chylomicrons 824
chyme 811
defecation 796, 838
deglutition 807
digestion 796
digestive tract 796

duodenum 817
esophagus 806
feces 833
gallbladder 832
gastric juice 814
hard palate 803
intestinal juice 821
large intestine 833
liver 827
lower esophageal orifice 810
mesentery 809
mucosal glands 819
pancreas 825
pepsin 816
peristalsis 796, 807

peritoneum 809
pharynx 806
plicae circulares 819
porta hepatis 830
pyloric orifice 810
rectum 835
salivary amylase 800, 803
salivary glands 803
secretin 827
small intestine 817
soft palate 803
stomach 810
submucosal glands 819
villi 819

Post Test

Matching

___ 1. Smooth muscle layer in the intestinal wall
___ 2. A slippery membrane surrounding the GI tract
___ 3. A glandular epithelium
___ 4. Goodbye mouth, hello pharynx!
___ 5. Richly supplied with blood vessels and nerves
___ 6. Bounded by cheeks outside and by teeth and gums inside
___ 7. Makes it difficult to swallow your tongue
___ 8. This one can make you gag

a. mucosa
b. submucosa
c. muscularis
d. serosa
e. lingual frenulum
f. buccal cavity
g. fauces

___ 9. Absorbs minerals and water, and removes waste
___ 10. Mastication-digestive process begins
___ 11. Major site of hydrolysis and absorption of nutrients.
___ 12. Initiates swallowing
___ 13. Conveys food to the stomach
___ 14. Stores food, initiates protein digestion
___ 15. First to receive salivary amylase.

a. mouth
b. esophagus
c. stomach
d. small intestine
e. large intestine
f. pharynx

Answer the following questions in the spaces provided.

16. How many cusps does a premolar tooth have? ___

17. What is another name for the eyeteeth? ___

18. What are the cutting teeth? ___

19. What holds the teeth in their sockets? ___

20. How many teeth make up a full adult dentition? ___

Identify the structures indicated in the diagram below.
___ 21. enamel
___ 22. pulp
___ 23. dentine
___ 24. cementum
___ 25. crown
___ 26. root canal

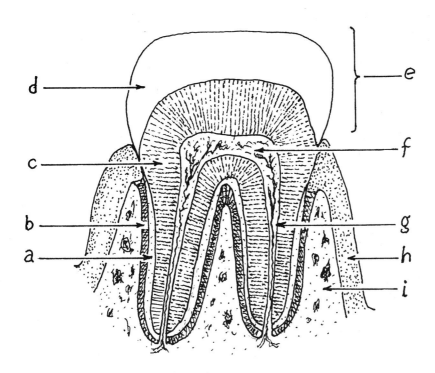

Matching

___ 27. Helps maintain pH of saliva between 6.5 and 6.85
___ 28. Lubricates food, helps form the bolus
___ 29. Hydrolyses carbohydrate
___ 30. Dissolves and softens food, moistens mouth, facilitates taste
___ 31. Important bacteriocidal function
___ 32. A salivary excretion
___ 33. Activates salivary amylase
___ 34. Part of salivary component of the immune system

a. IgA
b. HCO_3^-, PO_4^{3-}
c. H_2O
d. Cl^-
e. lysozyme
f. mucus
g. salivary amylase
h. urea

Short Answer

35. What has taste buds, three types of papillae, and several bands of interlacing muscles?

_____

36. What digestive secretion is entirely under nervous control? _____

37. What is the name for the sequential contractions of a tubular organ that proceed as a wave? _____

38. Identify the folded membrane, often containing large amounts of fat. that extends from the stomach down into the pelvic cavity? _____

39. What structure regulates the movement of food from the stomach to the duodenum?

_____

40. What are the longitudinal folds of the gastric mucosa called? _____

Matching

___ 41. Secreted by the chief cells of the stomach

___ 42. Secreted by the enteroendocrine cells of the stomach

___ 43. Secreted by the surface gland cells of the stomach

___ 44. Secreted by the stomach's parietal cells

___ 45. Secreted by neck cells of the gastric pits

___ 46. Also secreted by the duodenum

a. mucus
b. pepsinogen
c. HCl
d. gastrin, cholecystokinin, somatostatin

Matching

___ 47. A neural response that decreases gastric motility and gastric secretion

___ 48. Activated outside its source cell by HCl

___ 49. Formed by combining a component of gastric juice with Vitamin B_{12}

___ 50. Inhibits gastric emptying

___ 51. The mechanism that forces chyme out of the stomach

a. enterogastric reflex
b. pyloric pump
c. antianemic factor
d. pepsinogen
e. CCK, secretin, and GIP

Matching

____ 52. Intestinal glands(Crypts of Lieberkühn)
____ 53. Ventral lacteals
____ 54. Inward projections of lining epithelium
____ 55. They form the brush border of epithelial cells
____ 56. Lie deep within the intestinal glands
____ 57. Extend into the duodenal submucosa
____ 58. They occur throughout the GI tract, as well as in the ducts of the pancreas and liver and in the respiratory passages
____ 59. Form a system of internal spiraling folds
____ 60. Include a rich capillary network, as well as lacteals
____ 61. They secrete lysozyme

a. intestinal glands
b. plicae circulares
c. intestinal villi
d. lymphatic vessels
e. microvilli
f. duodenal submucosal (Brunner's) glands
g. Paneth cells
h. enteroendocrine cells

62. Name six major categories of substances that are digested and/or absorbed by the small intestine.

a. _____

b. _____

c. _____

d. _____

e. _____

f. _____

63. Complete the table below by inserting the letter that corresponds to the appropriate item in the following lists. (Note: Some items may be applicable in more than one place, and some may not be applicable at all.)

Sources
a. duodenal mucosa
b. colon
c. stomach
d. liver

Secretory Stimulus
e. acid or peptides in duodenum
f. fats in duodenum
g. protein in stomach
h. vagus nerve

Major Functions
i. stimulates HCl secretion
j. stimulates pancreas to secrete
k. stimulates release of bile
l. stimulates pepsinogen
m. inhibits HCl secretion
n. inhibits gastric/duodenal motility

Hormone	Source	Stimulus for Secretion	Major Function(s)
CCK			
GIP			
Gastrin			
Secretin			

64. What major blood vessels connect with the liver?

a. _____

b. _____

c. _____

d. _____

Matching

___ 65. Deamination
___ 66. Vitamins A ,D, E, and K
___ 67. Glucose → glycogen
___ 68. Glycogen → glucose
___ 69. Beta oxidation and ketogenesis

a. carbohydrate metabolism
b. lipid metabolism
c. protein metabolism
d. storage

Label the diagram below, using the letters for reference.
___ 70. cystic duct
___ 71. common bile duct
___ 72. hepatic duct
___ 73. hepaticopancreatic ampulla
___ 74. main pancreatic duct
___ 75. common bile duct sphincter

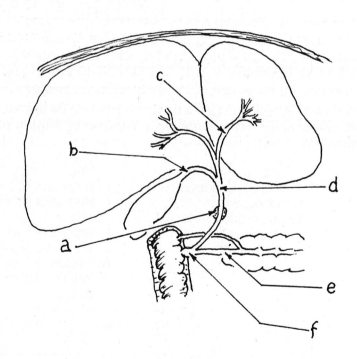

Integrative Thinking

1. Surgical removal of the entire stomach is done in cases of malignant cancer of that organ. Connecting the duodenum directly to the esophagus (duodenal-esophageal anastomosis) follows. If you were the attending physician for a patient who had been subject to this surgery, what instructions would you give to your patient with regard to diet and postoperative care, and for what reasons?

2. Baby Cynthia, a newborn, cried constantly and refused to nurse. When force-fed, she vomited the milk and cried with even greater distress. Unable to keep any nourishment down, she rapidly lost weight and had to be fed intravenously. Although she had an esophagus and a stomach and the rest of the GI tract, visualization by magnetic resonance imaging (MRI) revealed that her stomach contained some evidence of food following force-feeding but her small intestines were entirely empty. What would your diagnosis be?

3. What are the dangers of anorexia nervosa, other than starvation?

Your Turn

In Chapter 24 you learned how to make a concept map. Get your study group together and see where making a concept map that starts with the gallbladder will lead you. A few suggestions might help. Radiating above, to the right, and downward you can list the anatomical connections to the gallbladder (such as the liver and the pancreas) and then on the opposite side, the left, the physiological connections of the gallbladder to other functions and organs. You must exercise some restraint, however, because if you don't, you will find that everything in the body is connected in some way, directly or indirectly, to everything else in the whole body. So keep your connections direct and immediate. In subsequent chapters you may be asked to combine anatomical and physiological components in more sophisticated ways that will challenge your ingenuity, as well as your understanding. Bonne chance!

26 Metabolism, Nutrition, and the Regulation of Body Heat

Active Reading

Introduction

1. What are the six kinds of essential nutrients?

    a. _____

    b. _____

    c. _____

    d. _____

    e. _____

    f. _____

2. In what form does the energy in food exist? _____

3. What is cellular respiration? _____

I.   Review of Chemical Reactions (852)

1. Identify the following processes as either anabolic or catabolic.

    a.   biosynthesis _____

    b.   reaction releasing heat _____

    c.   macromolecule synthesis _____

    d.   reactions requiring energy input _____

    e. cellular respiration _____

    f. protein synthesis _____

    g. digestion _____

2. Contrast oxidation and reduction. _____

3. What is the common use for the energy in the reduced forms of the nucleotides NAD$^+$ and FAD? _____

II. Introduction to Carbohydrate Metabolism (852)

1. What are the four stages of carbohydrate catabolism?

    a. _____

    b. _____

    c. _____

    d. _____

III. Carbohydrate Metabolism (852)

1. What does the word root lysis mean? _____

2. What is phosphorylation? _____

3. How many ATP molecules are required for the glycolysis reactions to take place? _____. How many ATP molecules do the glycolysis reactions produce? _____. What is the net gain of ATP during glycolysis? _____.

4. The production of what molecules marks the end of the glycolysis stage? _____

5. How many CO_2 molecules are released from each pyruvic acid molecule prior to the citric acid (Krebs) cycle? _____

6. How many CO_2 molecules are released during the citric acid (Krebs) cycle? _____

7. What is the biologically useful product of cellular respiration? _____

8. How many molecules of ATP can be formed by the passage of electrons from each NADH + H$^+$ along the electron transport system? _____

9. How many molecules of ATP are produced from each molecule of glucose by means of the electron transport system? _____

10. What two transport mechanisms are involved in the electron transport system?

    a. _____

    b. _____

11. Where is glycogen stored? _____

12. What is the storage form of excess glucose that cannot be stored as glycogen? _____

13. What are three noncarbohydrate sources of glucose?

    a. _____

    b. _____

    c. _____

IV. Protein Metabolism (862)

1. What are the five uses for amino acids?

   a. _____

   b. _____

   c. _____

   d. _____

   e. _____

2. In the space below, write the chemical equation for oxidative deamination.

3. What happens to the leftover ammonia from oxidative deamination? _____

4. What does protein catabolism come at the expense of? _____

5. What is the difference between essential and nonessential amino acids? _____

   _____

6. By what process are nonessential amino acids formed in the body? _____

7. To what kind of molecule are most amino groups transferred during the process identified above? _____

8. What four hormones regulate protein synthesis?

   a. _____

   b. _____

   c. _____

   d. _____

V. Lipid Metabolism (864)

1. What are the most abundant lipids? _____

____ 2. Which of the following compounds is <u>not</u> a lipid?
   a. fatty acids
   b. sterols
   c. most alcohols
   d. carbohydrates
   e. fat-soluble vitamins

____ 3. What property do all lipids share?
   a. all are made up of fat
   b. all are stored in the same tissue
   c. all have common solubility properties
   d. all are exclusively harmful to body
   e. all of the above

4. Into what two substances are digested fats broken down?

   a. _____

   b. _____

5. What sequence of four reactions takes place in the mitochondrial matrix?

    a. _____

    b. _____

    c. _____

    d. _____

6. What are ketone bodies? _____

7. What is the prime determinant of the amount of lipid deposited in the adipocytes? _____

    _____

VI. Absorptive and Postabsorptive States (867)

1. What is the body's main energy source during the absorptive state?_____

2. What are chylomicrons? _____

3. What supplies most of the liver's energy during the absorptive state? _____

4. What three mechanisms does the body use to maintain a constant glucose concentration during the postabsorptive state?

    a. _____

    b. _____

    c. _____

5. How long after a heavy meal does the postabsorptive state take place? _____

6. What does the body use as additional sources of glucose during long periods of fasting?

    _____

7. Why do most body cells spare their use of glucose during the postabsorptive state?

    _____

VII. Nutritional Needs and Balance (869)

1. What are the six classes of essential nutrients?

    a. _____

    b. _____

    c. _____

    d. _____

    e. _____

    f. _____

2. What is the difference between a calorie and a Calorie?_____

3. From what sources do most carbohydrates come? _____

_____ 4. Which of the following is <u>not</u> a source of protein?

    a. animal fat
    b. beans
    c. red and white meats
    d. nuts
    e. fish

5. Compare the amounts of energy stored in proteins, carbohydrates, and fats._____
   _____

6. How much protein does a person need? _____

7. What leads to obesity? _____

8. What is the visible difference between saturated and unsaturated fats?_____

9. What two vitamins are produced in the body?

   a. _____

   b. _____

10. What are the fat-soluble vitamins?

   a. _____

   b. _____

   c. _____

   d. _____

11. What are the water-soluble vitamins?

   a. _____

   b. _____

12. Name seven macrominerals.

   a. _____

   b. _____

   c. _____

   d. _____

   e. _____

   f. _____

   g. _____

13. Complete the following table.

Mineral	Major Sources	Functions
Calcium		
Chloride		
Magnesium		
Phosphorus		
Potassium		
Sodium		
Sulfur		

14. What is the most important nutrient?_____

VIII.   Metabolic Rate and Temperature Control (874)

1. How does the body make use of the heat by-product of metabolism?_____

2. For each of the following factors, indicate whether it raises or lowers the metabolic rate.

   a. eating a protein-rich meal _____

   b. vigorous exercise _____

   c. increased epinephrine secretion _____

   d. decreased thyroxine secretion _____

   e. depression _____

   f. anxiety _____

   g. fear _____

   h. sexual activity _____

   i. aging _____

   j. pregnancy _____

   k. fever _____

   l. dieting without exercise _____

3. What are the units by which BMR is expressed? _____

4. How much heat is released for every liter of oxygen consumed? _____

Match the following pairs.

___ 5. Takes place not only on the skin surface,   a. radiation
       but also in the lungs   b. conduction
___ 6. Accounts for 50–60% of heat loss in a   c. convection
       normally heated room   d. evaporation
___ 7. Accounts for only about 3% of heat loss
___ 8. Accounted for by wind-chill factors

9. What are the three major components of the negative feedback system regulating body temperature?

   a. _____

   b. _____

   c. _____

10. Contrast core temperature and surface temperature._____

11. What is the hypothalamic thermostat? _____

12. What are the three ways by which the hypothalamic thermostat reduces body temperature?

   a. _____

   b. _____

   c. _____

13. What are the three ways by which the hypothalamic thermostat raises body temperature?

    a. _____

    b. _____

    c. _____

14. What is the function of brown fat? _____

IX. When Things Go Wrong (881)

1. What is the definition of obesity? _____

2. What are the visible symptoms of kwashiorkor? _____

3. What symptoms of hypervitaminosis are associated with each of the following vitamins?

    a. A _____

    b. D _____

    c. K _____

4. What enzymes are deficient in cases of cystic fibrosis?

    a. _____

    b. _____

    c. _____

5. What can be done to control the effects of PKU? _____

6. What can be done to prevent Tay-Sachs disease? _____

7. What is the appropriate immediate response to frostbite? _____

8. What is the treatment for heat prostration? _____

9. What are the possible effects of heatstroke? _____

10. What are the symptoms of hypothermia? _____

11. What is the dive reflex? _____

Key Terms

absorptive state 867
anabolism 852
basal metabolic rate 875
beta oxidation 864
calorie 869
catabolism 852
cellular respiration 851, 853

chemiosmosis 858
citric acid cycle 852, 853
electron transport system 852, 856
essential amino acids 864
gluconeogenesis 862
glycogenesis 861
glycogenolysis 862

glycolysis 852, 853
metabolic rate 874
metabolism 851
nonessential amino acids 864
oxidative deamination 862
postabsorptive state 869
triglycerides 864

Post
Test

Matching

___ 1. The process whereby electrons are given up by an atom or molecule

___ 2. The process whereby electrons are gained by an atom or molecule

___ 3. The chemical and energy transformations characteristic of living organisms

___ 4. The process by which nutrients are mechanically and/or chemically reduced in size and complexity and altered in solubility

___ 5. The process whereby relatively complex molecules are reduced chemically to smaller or less complex chemical components

___ 6. The building up of smaller molecules to larger and more complex molecules

___ 7. The process wherein a glucose molecule is chemically split to yield two molecules of pyruvic acid

___ 8. Biosynthesis

___ 9. An example is cellular respiration

___ 10. An example is FAD → FADH$_2$

a. metabolism
b. anabolism
c. catabolism
d. oxidation
e. reduction
f. digestion
g. glycolysis

Multiple Choice

___ 11. The most universally used molecule in energy transformations

a. oxygen
b. hydrogen
c. protein
d. carbohydrate
e. ATP

___ 12. The source of energy used to activate the breakdown of glucose

a. H$_2$O
b. O$_2$
c. ATP
d. NAD$^+$
e. FAD

___ 13. The molecule produced by glycolysis

a. glucose
b. pyruvic acid
c. ATP
d. citric acid
e. coenzyme A (CoA)

___ 14. The series of chemical transformations that occur in the mitochondria in which CO_2 and H^+ are produced

a. Krebs (citric acid) cycle
b. electron transport system
c. glycogenolysis
d. glycolysis
e. glycogenesis

___ 15. The series of stepwise oxidation-reduction reactions that occur in the inner membrane of mitochondria and generate ATP and H_2O

a. citric acid (Krebs) cycle
b. electron transport system
c. glycogenolysis
d. glycolysis
e. deamination

Matching

___ 16. The conversion of glucose to glycogen
___ 17. The production of glucose from a non-carbohydrate source, such as lactic acid or glycerol
___ 18. The conversion of glucose to pyruvic acid
___ 19. The conversion of glycogen to glucose
___ 20. A preliminary step in the conversion of protein into glucose

a. glycolysis
b. glycogenesis
c. glycogenolysis
d. glyconeogenesis
e. deamination

___ 21. Results in the production of ammonia
___ 22. Proteins that contain all nine amino acids
___ 23. Amino acids that cannot be synthesized by the human body and must be provided by the diet
___ 24. The 11 acids that can be synthesized by human metabolism
___ 25. A process by which one kind of amino acid can be converted to a different kind of amino acid

a. deamination
b. essential amino acid
c. nonessential amino acid
d. complete protein
e. transamination

___ 26. Speeds up the transport of amino acids into cells
___ 27. May be used as an energy source for protein synthesis
___ 28. Promote protein synthesis by increasing the efficiency of relevant genetic mechanisms
___ 29. Directly effects increased rate of protein synthesis
___ 30. May degrade proteins to be used for energy when other sources are inadequate

a. growth hormone (GH)
b. insulin
c. thyroxine
d. glucocorticoids
e. carbohydrates and lipids

Multiple Choice

____ 31. The most abundant lipids

    a. ketones
    b. fat-soluble vitamins
    c. triglycerides
    d. lipoproteins
    e. steroid hormones

____ 32. The products of hydrolysis of neutral fats

    a. glycerol
    b. fatty acids
    c. pyruvic acid
    d. both a and b
    e. none of the above

____ 33. Cells that synthesize and store triglycerides

    a. adipocytes
    b. monocytes
    c. carotenoids
    d. lipocytes
    e. beta choanocytes

____ 34. Substances derive from acetyl-CoA are collectively called

    a. acetyl-CoB's
    b. pyruvates
    c. acetone
    d. ketone bodies
    e. acetyl ketogens

____ 35. Nutrients are transferred from the digestive tract into the blood and lymph during the

    a. preabsorptive state
    b. absorptive state
    c. postabsorptive state
    d. fasting state
    e. chylomicron state

Matching

____ 36. The source of most of the body's energy supply during the postabsorptive state

____ 37. Small lipid droplets synthesized in the intestinal mucosa

____ 38. Molecules released into the blood during fasting to be used by tissue cells to produce CO_2, H_2O, and ATP.

____ 39. The shift of metabolism during fasting to utilize fat instead of glucose so that the latter can be conserved for the nervous system

____ 40. Synthesized in the liver from amino acids by removal of their NH_2 groups during the absorptive state

    a. chylomicrons
    b. keto acids
    c. stored fat
    d. ketone bodies
    e. glucose sparing

344 **Chapter 26**

_____ 41. The amount of energy required to
raise the temperature of one kilo-
gram of water one degree Celsius

    a. one calorie
    b. ten calories
    c. one Calorie
    d. ten Calories
    e. one thousand calories

_____ 42. With few exceptions, the source of
dietary carbohydrates is

    a. plants in general
    b. bread in particular
    c. pasta especially
    d. potatoes: baked, boiled, or fried
    e. none of the above

_____ 43. A carbohydrate that is indigestible
for human beings, but otherwise a
beneficial dietary substance

    a. glucose
    b. saccharose
    c. sucrose
    d. fructose
    e. cellulose

_____ 44. The most concentrated source of
energy in foods

    a. carbohydrates
    b. proteins
    c. fats
    d. ATP
    e. galactosides

_____ 45. Valuable plant sources for dietary
proteins are

    a. lilies (onions, leeks, garlic)
    b. legumes (beans, soybeans, peas, peanuts)
    c. cereals (wheat, barley, rice, oats)
    d. nuts (walnuts, cashews, almonds)
    e. all of the above except a

46. Name seven major dietary minerals.

    a. _____

    b. _____

    c. _____

    d. _____

    e. _____

    f. _____

    g. _____

47. Name seven dietary microminerals.

    a. _____

    b. _____

    c. _____

    d. _____

    e. _____

    f. _____

    g. _____

Multiple Choice

___ 48. Which of the following phenomena is a mechanism by which the hypothalamus "thermostat" reduces body heat?

    a. increase in glyconeogenesis
    b. decrease in glyconeogenesis
    c. constriction of cutaneous (skin) blood vessels
    d. dilation of cutaneous blood vessels
    e. shivering

___ 49. Which of the following is a hypothalamus-mediated mechanism for increasing body heat?

    a. constriction of cutaneous blood vessels
    b. dilation of cutaneous blood vessels
    c. inhibition of shivering
    d. relaxation of arrector pili muscles
    e. stimulation of glyconeogenesis

Integrative Thinking

1. Hundreds of years ago, mariners learned that being too long at sea invariably resulted in scurvy among the crew. What is scurvy, how was it associated with long sea voyages, and how was it ultimately prevented?

2. If you had a patient who was suffering vitamin deficiency, but did not know of which one or ones, (a) what question would you think to ask the patient right at the start? (b) If the symptoms included scaly skin (eczema), insomnia, fatigue, tendency toward depression, and constipation, which vitamin or vitamins would you recommend to the patient?

3. What would you prescribe for a patient who complained of light sensitivity, sores around the eyes, splitting of the skin, cataracts, vomiting, diarrhea, and muscular spasticity?

Your Turn

Play detective! Search in the library for a detailed description of techniques for determining BMR, including such basic techniques for determining how the consumption of oxygen is measured and how body surface area is measured. Share your findings with the members of your study group, and together write them up. Be sure to include a bibliography in your report.

27 The Urinary System

Active Reading

Introduction

1. What are the two general functions of the urinary system?

    a. _____

    b. _____

2. What is excretion? _____

I. The Urinary System: Components and Functions (888)

1. What are the four components of the urinary system?

    a. _____

    b. _____

    c. _____

    d. _____

2. What are the functional units of the kidneys? _____

3. What three functions do they perform?

    a. _____

    b. _____

    c. _____

4. On the figure below, label the following: diaphragm, liver, adrenal gland, renal pelvis, kidneys, ureter, urinary bladder, urethra, renal artery, renal vein, abdominal aorta.

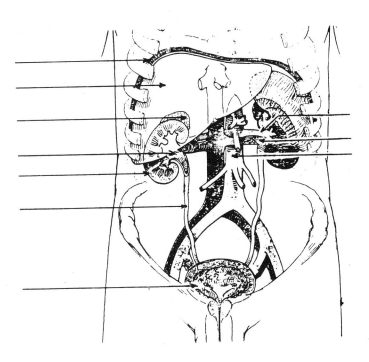

II. Anatomy of the Kidneys (888)

1. What percentage of water excreted from the body takes the form of urine? _____

2. Where do the artery, vein, and ureter enter and leave the kidney? _____

Match the following pairs.

___ 3. Innermost layer, continuous with the surface of the ureters

___ 4. Middle layers, composed of perirenal fat

___ 5. Outer layer covering and supporting kidney

a. adipose capsule
b. renal fascia
c. renal capsule

6. What happens to the kidneys with each breath? _____

7. On the figure below, identify the following: ureter, hilus, renal pelvis, renal artery, renal vein, major calyx, papilla, medulla, cortex, minor calyx, renal pyramid, renal column, nephron, renal fascia, adipose tissue, renal capsule, arcuate artery, interlobar vein, interlobular vein, arcuate vein, interlobar artery, interlobular artery.

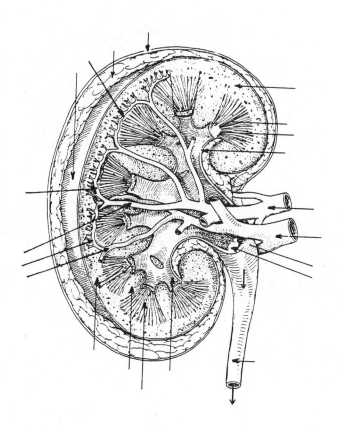

8. What percent of the cardiac output goes to the kidneys? _____

9. What is the site of blood filtration? _____

10. What is the second capillary bed? _____

11. Where do the many lymphatic vessels converge to form several large vessels? _____

12. How does the innervation of the kidneys regulate blood flow there? _____

13. On the figure below, identify the following: glomerular capsule, proximal convoluted tubule, descending loop, ascending loop, distal convoluted tubule, collecting duct, glomerulus, peritubular capillaries, vasa rectae.

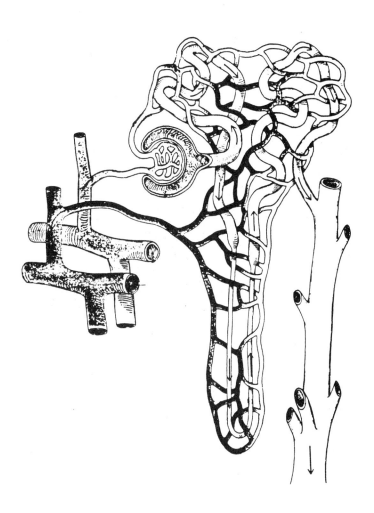

14. Complete the following table.

Structure	Reabsorption	Secretion
Proximal convoluted tubule		
Descending loop of nephron		
Ascending loop of nephron		
Distal convoluted tubule		
Collecting duct	Na^+, H_2O, K^+	H^+, K^+

15. Across what three layers does filtration of blood in the renal corpuscle take place?

    a. _____

    b. _____

    c. _____

16. What is the filtration barrier made up of? _____

17. What is glomerular filtrate? _____

18. True or false? The surfaces of the epithelial cells of the proximal convoluted tubule form a brush border on the surface of the lumen of the tubule. ____

19. True or false? Nephrons in the cortex have both a thick ascending limb and a thin ascending limb. ____

20. True or false? The cuboidal cells of the distal tubule contain more microvilli than those of the proximal tubule? ____

21. True or false? The juxtaglomerular apparatus helps to regulate blood pressure. ____

22. Of what is the juxtaglomerular apparatus composed? _____

III. Physiology of the Kidneys (896)

1. Describe the three processes by which the kidneys form glomerular filtrate and urine.

    a. _____

    b. _____

    c. _____

____ 2. The filtration in the renal corpuscle is driven by
    a. active transport
    b. sodium-potassium pumps
    c. hydrostatic pressure
    d. innervation
    e. none of the above

____ 3. The negatively charged glycoproteins in the basement layer of the capillaries hinders the passage of
    a. all macromolecules
    b. positively charged macromolecules
    c. no macromolecules
    d. negatively charged macromolecules
    e. none of the above

____ 4. The blood pressure in the glomerular capillaries is higher than in the rest of the body because
    a. renal afferent arterioles are short
    b. branches of afferent arterioles are straight
    c. efferent arterioles have a high resistance
    d. secret second heart from Mars ("Elvis Organ") contributes to high pressure
    e. almost all of the above

5. What three factors contribute to the determination of net filtration pressure?

    a. _____

    b. _____

    c. _____

6. What is the normal glomerular filtration fraction? _____

7. What is the filtration rate in a normal 70-kg male? _____

8. What percent of the glomerular filtrate is reabsorbed? _____

9. What are the two factors affecting GFR?

    a. _____

    b. _____

10. Describe how renal autoregulation responds to changes in blood pressure. _____

    _____

11. Describe the negative feedback system of the tubuloglomerular feedback mechanism.

    _____

12. Why would the extrinsic sympathetic control of GFR override the intrinsic renal
    autoregulatory control in situations of hemorrhaging? _____

    _____

13. What is tubular reabsorption? _____

14. Across what three membrane barriers do substances move as they are reabsorbed?

    a. _____

    b. _____

    c. _____

15. What two transport mechanisms are involved in reabsorption?

    a. _____

    b. _____

16. Complete the following table.

Substance	Amount Filtered by Kidneys Daily	Amount Reab-sorbed Daily	Percentage Reabsorbed Daily	Amount Excreted in Urine Daily
Glucose				
HCO_3^-				
Na^+				
Water				
Cl^-				
Uric acid				
K^+				
Total solute				
Urea				
Creatinine				

17. Why will glucose spill over into the urine, whereas sodium will not? _____

_____

18. By what mechanism does water reabsorption occur in the descending limb of the nephron? _____

19. What effect does ADH have on kidney function? _____

20. What substances are secreted in proximal and distal tubules and in the collecting duct?

    a. _____

    b. _____

    c. _____

    d. _____

21. What is the importance of hydrogen-ion secretion? _____

22. What are the factors regulating the active secretion of potassium ions?

    a. _____

    b. _____

23. What is plasma clearance? _____

24. What three general rules apply to all plasma clearance?

    a. _____

    b. _____

    c. _____

25. On what five factors does the countercurrent mechanism depend?

    a. _____

    b. _____

    c. _____

    d. _____

    e. _____

IV. Accessory Excretory Structures (906)

1. What are the three tissue layers of the ureters?

    a. _____

    b. _____

    c. _____

2. Where do the ureters transport urine? _____

3. What are the three main layers of the urinary bladder?

    a. _____

    b. _____

    c. _____

4. On the figure below, label the following: ureter, orifice of ureter, urethral orifice, internal smooth muscle sphincter, external urethral sphincter, urethra, tunica muscularis, urogenital diaphragm.

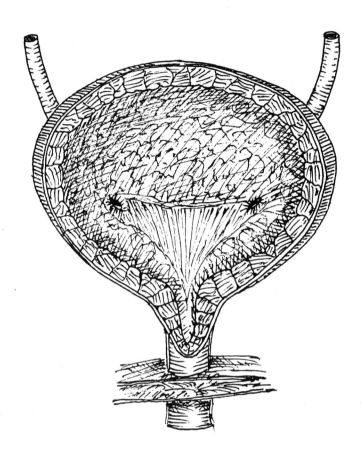

5. What is the trigone? _____

6. What type of muscle tissue makes up the external urethral sphincter? _____

7. What are the three portions of the male urethra?

   a. _____

   b. _____

   c. _____

V. Urine and Urination (911)

1. What is the average percent of water in urine?_____

2. What are two other normal constituents of urine?

   a. _____

   b. _____

3. What condition does the presence of protein in the urine indicate?_____

4. What condition does glucose in the urine indicate? _____

5. What are casts? _____

6. How does the pH of the urine of a meat eater differ from that of a vegetarian? _____

_____

7. What is the normal specific gravity range of urine? _____

8. Why is urine excretion reduced in hot weather? _____

9. What is the mechanism by which diuretics increase urine excretion? _____

10. Which autonomic nervous response (sympathetic or parasympathetic) does the micturition reflex involve? _____

11. How does the micturition reflex operate? _____

VI. The Effects of Aging on the Urinary System (912)

1. How much of its original tissue does each kidney need in order to function properly?

2. What causes incontinence? _____

3. What effect on urination does enlargement of the prostate gland have in elderly men?

_____

VII. When Things Go Wrong (912)

1. What are some causes of acute renal failure? _____

2. What is the treatment for chronic renal failure?

   a. _____

   b. _____

3. What are the symptoms of acute glomerulonephritis? _____

4. What causes acute pyelonephritis? _____

5. At what three sites may kidney stones become lodged?

   a. _____

   b. _____

   c. _____

6. From what does the most common type of renal calculus form? _____

7. What are the usual symptoms of a urinary tract infection? _____

8. What environmental factors are implicated in the development of cancer of the urinary
bladder? _____

9. Speculate as to why truck drivers and riders of horses and motorcycles might be susceptible to nephroptosis. _____

10. Describe some of the congenital abnormalities of the urinary system. _____

Key Terms

calyces 890
collecting duct 896
countercurrent multiplier system 905
distal convoluted tubule 896
glomerular capsule 894
glomerular filtrate 896, 898
glomerular filtration 897, 898
glomerular filtration rate 898
glomerulus 893
juxtaglomerular apparatus 896
loop of the nephron 896

micturition 912
nephron 894
net filtration pressure 898
peritubular capillaries 893
plasma clearance 904
proximal convoluted tubule 896
renal artery 893
renal corpuscle 894
renal cortex 893
renal medulla 890

renal pelvis 890
renal vein 893
tubular reabsorption 897, 901
ureter 907
urethra 909
urinary bladder 907
urinary system 888
urination 912
urine 911
vasa recta 894

Post Test

Multiple Choice

____ 1. Which of the following organs play a role in excretion?

    a. kidney
    b. lung
    c. skin
    d. salivary glands
    e. all of the above

____ 2. What biological function does excretion serve?

    a. homeostasis
    b. homeostasis
    c. homeostasis
    d. all of the above

____ 3. The large, collecting space within the kidney is the

    a. major calyx
    b. renal medulla
    c. renal pelvis
    d. renal cortex
    e. renal pyramid

____ 4. The renal medulla consists of several

    a. major calyces
    b. minor calyces
    c. renal pelvices
    d. renal cortices
    e. renal pyramids

___ 5. The granular-appearing, outermost
portion of the kidney is the renal

a. cortex
b. medulla
c. pyramid
d. calyx
e. column

___ 6. The anatomical unit of kidney
function is the

a. cortex
b. medulla
c. nephron
d. collecting tubule
e. glomerulus

7. Trace the course of blood through
the kidney, using the following
lettered list.

( )( )( )( )( )( )( )( )( )( )

a. peritubular capillaries
b. arcuate arteries
c. arcuate veins
d. efferent arterioles
e. afferent arterioles
f. renal artery
g. renal vein
h. interlobular arteries
i. interlobular veins
j. interlobar arteries
k. interlobar veins
l. glomerulus

8. In a manner similar to that of the
preceding question, trace the route
of filtrate through a nephron.

( )( )( )( )( )( )( )( )

a. glomerulus
b. distal convoluted tubule
c. proximal convoluted tubule
d. collecting duct
e. podocytes
f. macula densa
g. loop of the nephron (Henle)
h. filtration slits

9. Where, in relation to the nephric tubules, are the vasa recta located? _____

Matching

___ 10. The fluid that enters the proximal
convoluted tubule
___ 11. Specialized smooth muscle cells
___ 12. Supplies the sympathetic innervation
of the kidney
___ 13. Specialized epithelial cells that
surround the glomerular capillaries
___ 14. The contribution of the distal convoluted tubules to the JGA
___ 15. Form pedicels

a. juxtaglomerula cells
b. podocytes
c. glomerular filtrate
d. macula densa
e. renal plexus

Identify the anatomical structures indicated in the figure below.

___ 16. juxtaglomerular apparatus
___ 17. glomerulus
___ 18. proximal convoluted tubule
___ 19. distal convoluted tubule
___ 20. descending loop of the nephron
___ 21. ascending loop of the nephron
___ 22. collecting duct
___ 23. afferent arteriole
___ 24. efferent arteriole

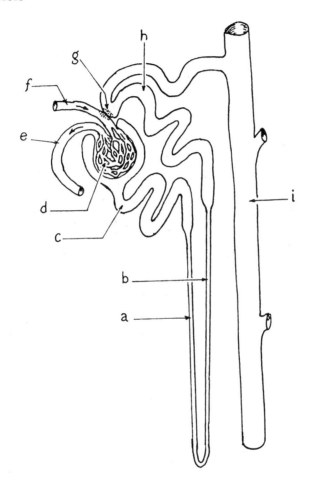

Correlate the structures indicated in the figure above with the following function(s).

___ 25. Conduction of urine to the renal pelvis.
___ 26. Reabsorption of HPO_4^{2-}, Cl^-, SO_4^{2-}, HCO_3^-.
___ 27. By hydrostatic pressure, filtration of most dissolved substances from the blood (excepting blood cells and most plasma proteins).
___ 28. Reabsorption of Na^+ by active transport, water by osmosis, and urea by diffusion.
___ 29. Reabsorption of Na^+ and Cl^- by active transport, and of HCO_3^- by electrochemical gradient.
___ 30. Monitoring and helping regulate systemic blood pressure, among other things.

Identify the structures indicated in the figure below.

___ 31. urinary bladder
___ 32. penile urethra
___ 33. voluntary (external) sphincter
___ 34. ureter
___ 35. involuntary (internal) sphincter

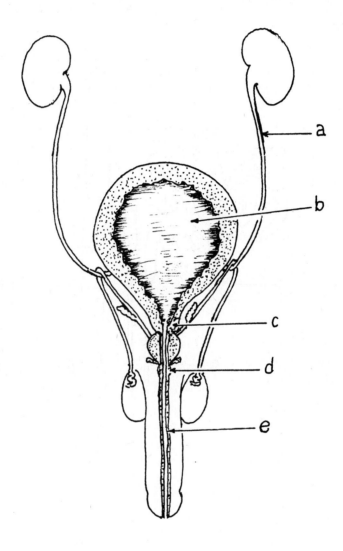

Matching

___ 36. Urine cloudy, owing to presence of fat droplets or pus
___ 37. Urine containing protein
___ 38. The act of urination
___ 39. Presence of sugar in the urine
___ 40. Kidney stones
___ 41. A substance that increases urine production

a. glycosuria
b. pyuria
c. micturition
d. urinary calculi
e. diuretic
f. albuminuria

Integrative Thinking

1. Acute renal failure seems to have a significantly higher incidence among farmers, house painters, and automobile repair workers than among the general population. What do these occupations have in common? Do you think that acute renal failure can be considered to be an occupational disease?

2. A young woman went to her physician complaining of persistent back pain, feeling a little feverish, and that her urine was cloudy and had the odor of ammonia. The doctor learned that she also had dysuria, and lab tests revealed an elevated leukocyte count and bacteria in her urine. What is your diagnosis?

3. Space travelers visiting the distant planet Dune, learned that the inhabitants of that waterless planet had, somehow, adapted to the point that they were never thirsty, and the only water they ever ingested was whatever happened to be chemically derived from their food. If a pathologist were to examine the microscopic anatomy of the kidney of a deceased Dunsian, what do you think would be found different about the nephrons?

Your Turn

Try another concept map, using the following organs and associated structures: kidney, ureter, bladder, internal and external urinary sphincters, JG cells, macula densa, renal plexus, parasympathetic innervation, neurohypophysis, adrenal cortex, and the male reproductive system.

28 Regulation of Body Fluids, Electrolytes, and Acid-Base Balance

Active Reading

Introduction

1. What percent of one's total body weight is water?

2. What is the most serious effect of a loss of 20% of one's total body water?

I. Body Fluids: Compartments and Composition (920)

1. What percentage, by volume, of total body water is intracellular fluids? _____

2. Identify seven kinds of transcellular fluids.

    a. _____

    b. _____

    c. _____

    d. _____

    e. _____

    f. _____

    g. _____

3. What is interstitial fluid? _____

4. Complete the following table.

Substance	Concentration in Blood Plasma	Concentration in Interstitial Fluid	Concentration in Intracellular Fluid
Sodium			
Potassium			
Calcium			
Magnesium			
Bicarbonate			
Chloride			
Sulfate			
Protein			
Phosphate			

II. Movement of Fluids Between Compartments (921)

1. What two forces control the movement of water between body compartments?

   a. _____

   b. _____

2. Where in the body does the fluid in interstitial spaces return to the blood plasma? _____

3. What is the primary force regulating fluid flow between the inside of cells and the interstitial spaces? _____

4. What is the physiological effect of increased sodium concentration in the interstitial fluid? _____

III. Water (922)

1. Explain why a water imbalance will have exaggerated effects in an infant, compared with an adult. _____

2. Why do men have more water as a percentage of total body weight, on average, than women? _____

3. Name four physiological functions of water.

   a. _____

   b. _____

   c. _____

   d. _____

4. List the body's three sources of water, together with their percent contribution to the total.

   a. _____

   b. _____

   c. _____

5. What are the four components of the thirst mechanism?

    a. _____

    b. _____

    c. _____

    d. _____

6. What percent of total excreted water does each of the following organs account for?

    a. kidneys _____

    b. skin _____

    c. GI tract _____

    d. lungs _____

IV. Function and Regulation of Specific Electrolytes (925)

1. What are electrolytes? _____

2. What distinguishes anions from cations? _____

3. Name the physiologically most important electrolytes. _____

4. Identify four major functions of electrolytes in the body.

    a. _____

    b. _____

    c. _____

    d. _____

5. How are electrolytes lost from the body? _____

6. Complete the following table.

Hormone	Source of Secretion	Mode and Site of Action
	Heart	
	Adrenal glands	
	Hypothalamus; released by neurohypophysis	
	Parathyroid glands	

7. What three hormones regulate sodium levels in the plasma?

    a. _____

    b. _____

    c. _____

8. Which is more important in osmotic regulation, sodium or potassium? _____

9. Where is potassium concentration highest? _____

10. What generally dictates the movement of chloride ions? _____

11. What two hormones regulate the levels of calcium?

     a. _____

     b. _____

12. What are the three effects of PTH on calcium?

     a. _____

     b. _____

     c. _____

13. What two hormones regulate phosphate ion concentration?

     a. _____

     b. _____

14. What important role does magnesium play in the economy of the body? _____

V. Acid-Base Balance (927)

1. What is the pH of blood and the interstitial fluids? _____

2. What three mechanisms regulate the pH of extracellular fluids?

     a. _____

     b. _____

     c. _____

3. What is the normal ratio of bicarbonate to carbonic acid in the plasma? _____

4. What regulates carbonic acid concentrations? _____

5. What regulates bicarbonate ion concentrations? _____

6. Where does the phosphate buffer system regulate pH? _____

7. Complete the reaction

    $HCl + Na_2HPO_4 \rightarrow$

8. What is the fastest buffer system in the body? _____ Why is it the fastest? __

    _____

9. Explain the following reaction.

$$\underset{\text{base}}{R-\overset{\overset{\displaystyle H}{|}}{\underset{\underset{\displaystyle NH_2}{|}}{C}}-\overset{\overset{\displaystyle O}{\|}}{C}-OH} \cdots\cdots H^+ \rightarrow \underset{\text{acid}}{R-\overset{\overset{\displaystyle H}{|}}{\underset{\underset{\displaystyle NH_3^+}{|}}{C}}-\overset{\overset{\displaystyle O}{\|}}{C}-OH}$$

10. What is the effect of a decrease in CO_2 concentration? _____

11. Why is respiratory regulation of pH such an important short-term regulator? _____

_____

12. Give four reasons why the hydrogen-sodium exchange in the kidneys is significant.

    a. _____

    b. _____

    c. _____

    d. _____

13. What are the three ways in which the kidneys excrete hydrogen ions and thus regulate pH?

    a. _____

    b. _____

    c. _____

VI. When Things Go Wrong (932)

1. At what point is acidosis usually fatal? _____

2. What can cause respiratory acidosis? _____

3. What can cause respiratory alkalosis? _____

4. What are some signs of alkalosis? _____

5. Name five causes of edema.

    a. _____

    b. _____

    c. _____

    d. _____

    e. _____

6. What four pathological conditions can edema cause?

    a. _____

    b. _____

    c. _____

    d. _____

7. What are some signs of hyponatremia? _____

8. What is the kidneys' response to hypernatremia? _____

9. Within what range of potassium concentrations is one neither hypokalemic nor hyperkalemic? _____

Key Terms

acid 927
antidiuretic hormone 926
base 927
buffer system 928

electrolytes 925
extracellular fluid 920
hydrostatic pressure 921
intracellular fluid 920

osmotic pressure 921
parathyroid hormone 927

Post Test

Multiple Choice

_____ 1. The extracellular fluid compartment that includes cerebrospinal fluid and synovial fluid

a. interstitial fluid
b. transcellular fluid
c. tissue plasma
d. subcellular fluid
e. nonelectrolyte fluid

_____ 2. The forces primarily utilized in controlling water movement between body compartments

a. hydrostatic pressure
b. osmotic pressure
c. electrolyte dissociation
d. metabolic
e. both a and b

_____ 3. Location of a concentration of neuron cells that make up the thirst center

a. stellate ganglion
b. geniculate body
c. infundibulum
d. hypothalamus
e. Mike's Bar and Grille

_____ 4. What molecules that ionize in solution become

a. atoms
b. free radicals
c. enzymes
d. linkage groups
e. electrolytes

_____ 5. A low plasma sodium concentration

a. hypernatremia
b. hyponatremia
c. hyperkalemia
d. hypokalemia
e. hypocalcemia

_____ 6. What the concentration of electrolytes in the blood plasma is regulated by to a very considerable
extent

     a. diet
     b. the CNS
     c. hormones
     d. the GI tract
     e. the liver

_____ 7. An adrenocortical secretion that
regulates sodium reabsorption by the
nephric tubules

     a. ADH
     b. aldosterone
     c. cortisone
     d. oxytocin
     e. ANF

_____ 8. Water that is a by-product of the
oxidation of nutrients in the cells of
the body

     a. electrolyte transfer water
     b. water of hydration
     c. water of hydrolysis
     d. metabolic water
     e. expendable water

_____ 9. A decrease in sodium concentration
in the blood decreases the release of

     a. ADH
     b. oxytocin
     c. aldosterone
     d. ANF
     e. angiotensin

_____ 10. Controls potassium excretion

     a. liver
     b. adrenal gland
     c. heart
     d. kidney
     e. pituitary gland

Matching

_____ 11. Where sodium goes, this is sure to
follow — most of the time

_____ 12. Plays an important role in sodium-
potassium pumps, and in ATP
production in mitochondria

_____ 13. Under control of parathormone
(PTH) and calcitonin (CT); excessive concentration evident as
hypercalcemia

_____ 14. The concentration of this electrolyte
is also regulated by PTH and CT, but
it is most important for the high
energy bond it forms with adenosine
and for its central role in nucleic
acids

_____ 15. Most important in regulation of
osmosis and in propagation of nerve
impulses

_____ 16. Extremely important in generating
and maintaining resting membrane
potentials in nervous tissue

     a. Na^+
     b. Cl^-
     c. Ca^{2+}
     d. PO_4^{3-}
     e. K^+
     f. Mg^{2+}

Short Answer

17. Name the three acid-base buffer systems.

   a. _____

   b. _____

   c. _____

18. Which system is the most abundant in body cells?_____

19. What do buffers do?_____

20. Which buffer system is much the fastest?_____

21. How does the respiratory center in the medulla oblongata help to regulate the pH of the blood? _____

Multiple Choice

____ 22. Respiratory regulation of acid-base balance is primarily concerned with regulating

    a. phosphate buffers
    b. the dissociation of carbonic acid
    c. protein buffer systems
    d. hormonal regulation of pH
    e. renal regulation of pH

____ 23. The kidneys help to regulate plasma pH by

    a. secreting angiotensinogen
    b. excreting hydrogen ions
    c. excreting ammonium ions
    d. responding to ADH
    e. both b and c

____ 24. Renal excretion of H^+ is accompanied simultaneously by

    a. reabsorption of Na^+
    b. excretion of Na^+
    c. excretion of PO_4^{3-}
    d. reabsorption of NH_4^+
    e. reabsorption of H_2O

____ 25. An increase in the rate and depth of breathing increases the amount of CO_2 exhaled, which is accompanied by a reduction in the amount of ___ in the blood.

    a. carbonic acid
    b. H^+
    c. CO_2
    d. ammonia
    e. both a and b

368   Chapter 28

Matching

____ 26. May result from diabetes mellitis, malnutrition, or starvation

____ 27. Results from impaired ability to remove CO_2 from the lungs

____ 28. Results if the pH of the blood rises above 7.45 as a consequence of hyperventilation, high altitude, or emotional disturbances

____ 29. Manifested as excessive swelling of tissues, owing to an abnormal increase in interstitial fluid

____ 30. Results from an excessive loss of acid or uptake of alkaline substances. May be brought about by excessive vomiting of gastric hydrochloric acid

a. respiratory acidosis
b. metabolic acidosis
c. respiratory alkalosis
d. metabolic alkalosis
e. edema

Integrative Thinking

1. You have been hiking in the mountains for several days now, when your companion collapses from fatigue. But instead of resting quietly, she fidgets and is very restless. She munches on potato chips, at the same time complaining of extreme thirst. You took very little water with you, because it is so heavy, and neither of you has drunk very much as a result. Is your companion in real trouble? What should you do?

2. What are the causes of edema, and what pathological conditions might it lead to?

Your Turn

An Experiment: Materials: Get a small bottle of white vinegar (acetic acid) and some source of ammonia (ammonium hydroxide), such as certain glass cleaning preparations, some litmus papers for testing acid vs. base, two eyedroppers, and a small glass cup or saucer. Label one dropper for acid use and the other for base use. Procedure:

1. Test your acid and base ingredients with litmus paper. Wash the cup or saucer after each step in the experiment.

2. Put 10 drops of water and 10 drops of vinegar in the cup, using the acid dropper. Mix by agitation. Add ammonium hydroxide to the vinegar, drop-by-drop, with the base dropper. You may want to shake the test solution slightly to get good mixing. Test with litmus paper after each drop. How many drops of ammonia does it take to change the acidity? Does it change abruptly, slowly, or not at all? What is the significance of the results?

3. Repeat the reverse procedure, adding vinegar to the ammonia. Record your results.

4. Discuss the results among members of your group. Did the experiment work? What would you recommend to improve it? If it did not work, did you learn anything from it? What did you learn?

29 The Reproductive Systems

Active Reading

Introduction

1. What two sex cells does the reproductive system bring together?

   a. _____

   b. _____

2. What function does this serve? _____

I. Male Reproductive Anatomy and Physiology (941)

1. What are the four different types of structures involved in the male reproductive system?

   a. _____

   b. _____

   c. _____

   d. _____

2. On the figure below, identify the following: urinary bladder, ureter, testis, tunica albuginea, scrotum, epididymis, spermatic cord, seminal vesicle, ejaculatory duct, urethra, ductus deferens, prostate gland, bulbourethral gland, corpus cavernosum, corpus spongiosum, prepuce, glans penis, urethral orifice.

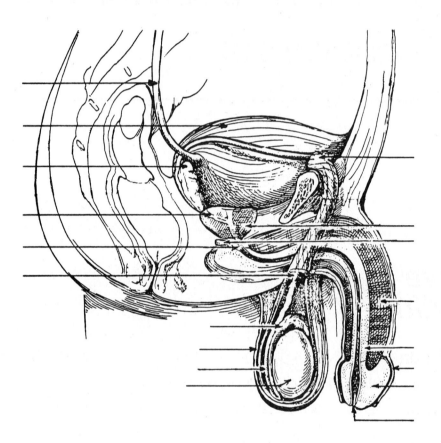

3. What is the fetal passageway leading to the scrotum? _____

4. How does the temperature of the testes compare with that of the rest of the body? _____

_____

5. What separates the two testes? _____

6. Where is the site of sperm production? _____

7. On the figure below, identify the following: tunica vaginalis, tunica albuginea, lobule, seminiferous tubule, efferent ducts, head of epididymis, body of epididymis, tail of epididymis, ductus deferens.

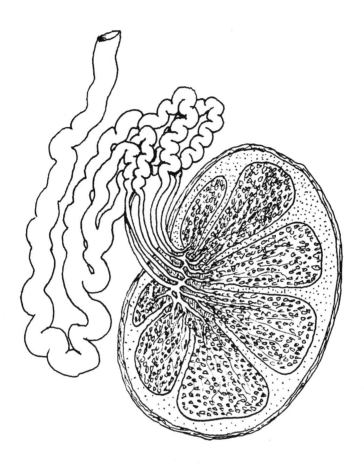

8. What four functions do sustentacular cells serve?

    a. _____

    b. _____

    c. _____

    d. _____

9. What is the most important androgen? _____

10. On the figure below, identify the following: head, acrosome, nucleus, end piece, flagellum, middle piece, tail, centrioles, mitochondria, plasma membrane.

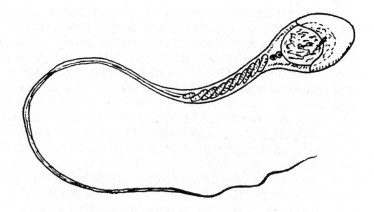

___ 11. The time required for a sperm to develop is somewhat more than

    a. 1 second
    b. 2.5 hours
    c. 28 days
    d. 2 months
    e. 14 years

___ 12. A normal ejaculation contains

    a. one sperm cell
    b. several thousand sperm cells
    c. about 1 million sperm cells
    d. 300 – 500 million sperm cells

13. What are the three main functions of the epididymis?

    a. _____

    b. _____

    c. _____

14. Where are sperm cells stored prior to ejaculation? _____

15. What is the function of the internal sphincter of the urethra? _____

16. What are the three sections of the urethra?

    a. _____

    b. _____

    c. _____

17. Complete the following table.

Gland	Location	Secretion	Time of Secretion
Bulbourethral glands			
		Prostaglandins, water, fructose, vitamin C	
			Continuously

18. What is the main component of semen? _____

19. Where are penile nerve endings most concentrated? _____

20. What two parts of the CNS control an erection?

    a. _____

    b. _____

21. What is detumescence? _____

22. What do the interstitial cells secrete?

    a. _____

    b. _____

    c. _____

23. Describe the means by which the hypothalamus controls the release of testosterone. ____

    _____

24. Besides spermatogenesis, what does testosterone control? _____

25. What is the role of inhibin in the maintenance of a proper sperm count? _____

    _____

II. Female Reproductive Anatomy and Physiology (950)

1. What are the six components of the female reproductive system?

    a. _____

    b. _____

    c. _____

    d. _____

    e. _____

    f. _____

2. On the figure below, identify the following: ovary, ureter, cervix, rectum, urethra, vagina, suspensory ligament, uterine tube, ovarian ligament, uterus, round ligament, urinary bladder, mons pubis, clitoris, labia majora and minora.

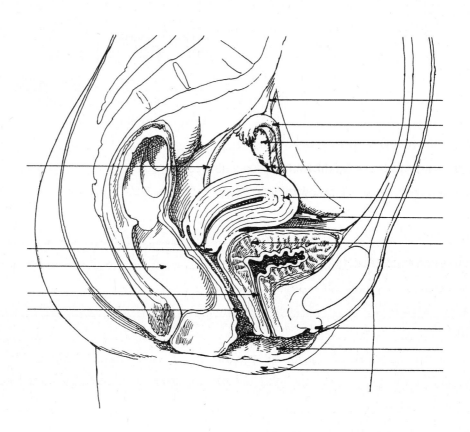

3. What layer covers each ovary? _____

Match the following pairs.

___ 4. Follicle that is not yet growing, with one layer of squamouslike follicular cells around the oocyte

___ 5. Temporary endocrine tissue derived from ruptured follicle

___ 6. Follicle with two or more layers of cuboidal or columnar cells around the oocyte

___ 7. Ready to release a secondary ovum

a. primary follicle
b. primordial follicle
c. vesicular follicle
d. corpus luteum

8. What two hormones does the corpus luteum secrete?

    a. _____

    b. _____

9. What is the response of the ovary to high levels of FSH and LH? _____

10. What happens to the number of primary oocytes during a person's lifetime? _____

11. How does the ovum move through the uterine tubes? _____

12. What are the three layers of the uterine tubes?

    a. _____

    b. _____

    c. _____

13. What is the function of the fimbriae? _____

14. What are the five ligaments, or types of ligaments, supporting the uterus?

    a. _____

    b. _____

    c. _____

    d. _____

    e. _____

Match the following pairs.

___ 15. Three layers of smooth muscle fibers, making up the bulk of the uterine wall

___ 16. Outer layer of uterus, extending to form two broad ligaments

___ 17. Mucous membrane, composed of the stratum functionalis and the stratum basalis

a. serosal layer of uterus
b. myometrium
c. endometrium

18. What does the breakup of the endometrium produce? _____

19. What are two effects of the acidity of the vagina?

    a. _____

    b. _____

20. Locate and describe the functions of the greater and lesser vestibular glands. _____

_____

21. Briefly describe the function of each of the following.

    a. areolar glands _____

    b. lactiferous ducts _____

    c. lactiferous sinus _____

III. The Menstrual Cycle (957)

1. Complete the following table.

Hormone	Source	Function
Estrogen		
FSH		
GnRH		
hCG		
LH		
Oxytocin		
Progesterone		
Prolactin		

2. What are the three ovarian phases of the menstrual cycle?

   a. _____

   b. _____

   c. _____

3. What are the three uterine phases of the menstrual cycle?

   a. _____

   b. _____

   c. _____

4. Complete the following table.

Day of Cycle	Ovarian Events	Uterine Events
1		
2		
3		
4		
5		
6		
7		
8		
9		
10		
11		
12		
13		
14		
15		
16		
17		
18		
19		
20		
21		
22		
23		
24		
25		
26		
27		
28		

IV. Formation of Sex Cells (Gametogenesis) (962)

Match the following pairs.

___ 1. General term for sex cell
___ 2. Specific type of nuclear division reducing number of chromosomes to half
___ 3. Union of female and male sex cells
___ 4. Formation of male sex cells
___ 5. Formation of female sex cells
___ 6. Diploid number
___ 7. Haploid number
___ 8. Paired homologs

a. gamete
b. fertilization
c. spermatogenesis
d. 46
e. tetrad
f. 23
g. meiosis
h. oogenesis

9. What is the effect of crossing-over? _____

10. Describe the six steps of spermatogenesis.

    a. _____
    b. _____
    c. _____
    d. _____
    e. _____
    f. _____

11. How many ova does the primary oocyte yield? _____

12. Describe the seven steps of oogenesis.

    a. _____
    b. _____
    c. _____
    d. _____
    e. _____
    f. _____
    g. _____

V. Conception (964)

1. For how long does the ovum remain viable? _____

2. For how long do the sperm remain viable? _____

3. What are two reasons for so many sperm cells being ejaculated at once?

    a. _____
    b. _____

4. What is the function of acrocin? _____

5. What is the zona pellucida? _____

6. Describe the four steps of penetration and fertilization.

    a. _____

    b. _____

    c. _____

    d. _____

7. What is a pronucleus? _____

8. How many autosomes does a diploid cell have? _____

9. At what point is the sex of a child determined? _____

VI. Contraception (968)

1. What are the five most effective forms of contraception?

    a. _____

    b. _____

    c. _____

    d. _____

    e. _____

2. What are the five least effective forms of contraception?

    a. _____

    b. _____

    c. _____

    d. _____

    e. _____

VII. The Effects of Aging on the Reproductive Systems (968)

1. At what age do symptoms of menopause usually appear? _____

2. What happens to levels of estrogen and testosterone with age? _____

3. What form(s) of cancer do women become increasingly susceptible to with age? _____

    _____

4. What form(s) of cancer do men become increasingly susceptible to with age? _____

    _____

VIII. Developmental Anatomy of the Reproductive Systems (970)

1. Until what age do the internal reproductive organs remain undifferentiated between males and females? _____

2. At what age do the external reproductive organs begin to differ between males and females? _____

3. At what age do the testes begin to descend into the scrotum? _____

IX. When Things Go Wrong (972)

1. What male organs does NGU infect?

   a. _____

   b. _____

   c. _____

2. What are the effects of NGU in women? _____

3. What are the symptoms of herpes II? _____

4. It has been said that, as opposed to love, herpes is forever. Why is this often true? _____

   _____

5. What are some symptoms of gonorrhea? _____

6. What is the treatment for gonorrhea? _____

7. To what long-term effects can syphilis lead?

   a. _____

   b. _____

   c. _____

8. What is the most common site of breast cancer? _____

9. What is the current favored treatment for breast cancer? _____

10. What is the most effective way of detecting cervical cancer? _____

11. What is the most common form of cancer among males? _____

12. What is BPH? _____

13. What is the treatment for ovarian cysts? _____

14. What are the symptoms of endometriosis? _____

Key Terms

acrosin 967
autosome 967
bulbourethral glands 947
corpora cavernosa 947
corpus luteum 950
corpus spongiosum 947
crossing-over 963
ductus deferens 945
endometrium 954
epididymis 945
follicle 950
FSH 949, 957
GnRH 949, 957

hCG 960
LH 949, 957
mammary glands 957
meiosis 962
menstruation 955
oogenesis 964
ovaries 950
ovulation 952, 960
oxytocin 962
polar body 964
primary oocyte 964
prolactin 962
prostate gland 946

scrotum 941
secondary oocyte 964
semen 947
seminal vesicles 946
seminiferous tubules 941
sex chromosomes 947
sperm 944
spermatogenesis 963
testes 941
uterine tubes 953
uterus 954
vagina 955

Post Test

Matching

____ 1. A white, fibrous sac that encloses the testis
____ 2. The condition of undescended testes
____ 3. Provides separate compartments for the two testes
____ 4. The epidermal sac that houses the testes
____ 5. The structure formed from the caudal genital ligament
____ 6. The tunnel through which the testes descend
____ 7. A tissue continuous with the peritoneal membrane

a. tunica vaginalis
b. tunica albuginea
c. scrotum
d. inguinal canal
e. cryptorganism
f. median septum
g. gubernaculum

Matching

____ 8. Site of sperm cell production
____ 9. Network of tubules that drain into the efferent ducts
____ 10. Secretes an androgen-binding protein
____ 11. Secretes testosterone
____ 12. The cells that produce the sperm cells

a. sustentacular (Sertoli) cell
b. spermatogenic cells
c. interstitial endocrinocyte
d. seminiferous tubules
e. rete testis

Label the diagram below.
___ 13. testis
___ 14. bulbourethral gland
___ 15. corpus cavernosum
___ 16. corpus spongiosum
___ 17. epididymis
___ 18. ejaculatory duct
___ 19. prostate gland
___ 20. seminal vesicle
___ 21. glans penis
___ 22. ductus deferens

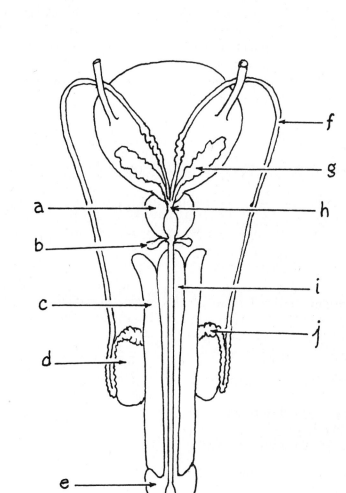

Matching

____ 23. Secreted by the hypothalamus
____ 24. Stimulates sustentacular cells in testes
____ 25. Released by sustentacular cells in response to elevated sperm cell count
____ 26. Stimulates interstitial cells of the testes
____ 27. Keeps testosterone from diffusing out of the seminiferous tubules
____ 28. Stimulates spermatogenesis
____ 29. Decreasing blood levels of this substance stimulate the hypothalamus
____ 30. Stimulates the testes to secrete androgens and a small amount of estrogen

a. follicle-stimulating hormone
b. luteinizing hormone
c. gonadotropin-releasing hormone
d. testosterone
e. inhibin
f. androgen-binding protein

____ 31. Stimulates thickening of uterine wall and maturation of oocytes
____ 32. Stimulates development of female sexual characteristics
____ 33. Stimulates uterine contractions of labor
____ 34. Stimulates thickening of uterine wall and formation of mammary ducts
____ 35. Stimulates further development of oocyte and follicle; aids development of the corpus luteum
____ 36. Maintains the corpus luteum
____ 37. Increases estrogen secretion and stimulates new oocyte formation
____ 38. Makes cervix more easily stretched and promotes softening of the symphysis pubis
____ 39. Promotes milk production
____ 40. Stimulates secretion of estrogen and progesterone by the corpus luteum

a. estrogen
b. progesterone
c. prolactin
d. FSH
e. LH
f. hCG
g. oxytocin
h. relaxin

Short Answer

41. What hormone(s) does the corpus luteum secrete? _____

42. What is the technical term for the production and development of egg cells? _____

43. What does the empty ovarian follicle become if pregnancy occurs? _____

44. What does the empty ovarian follicle become if pregnancy does not occur? _____

45. The endometrium is composed of two layers, a stratum basalis and a _____

Matching

___ 46. The ligament that attaches the uterus to the rectum

___ 47. Attaches the uterus to the urinary bladder

___ 48. Helps the uterus to stay tilted forward over the urinary bladder

___ 49. Extends from the uterus to the floor and lateral walls of the pelvic cavity

___ 50. The ligament that extends from the upper part of the cervix to the sacrum

a. broad ligament
b. round ligament
c. uterosacral ligament
d. posterior ligament
e. anterior ligament

Multiple Choice

___ 51. Diseases caused by microorganisms or infectious agents that are transmitted mainly by sexual contact are designated

a. STD
b. VD
c. NUG
d. GC
e. a or b

___ 52. The most common STD in the United States is

a. USP
b. UPS
c. RSVP
d. NGU
e. GC

___ 53. Nongonococcal urethritis is caused by

a. Trichomonas vaginalis
b. Chlamydia trachomatis
c. type I herpes simplex
d. type II herpes simplex
e. Treponema pallidum

___ 54. A protozoan organism that commonly infects the lower genito-urinary tract is

a. Trichomonas vaginalis
b. Chlamydia trachomatis
c. herpes simplex
d. Treponema pallidum
e. Neisseria gonorrhoeae

___ 55. Perhaps the most debilitating and deadly of the following STDs if left untreated is

a. gonorrhea
b. syphilis
c. type II herpes
d. chlamydia
e. trichomoniasis

Integrative Thinking

1. Ingrid was frightened. She was suffering almost constant pain in the lower back, occasional uterine bleeding, and her periods were irregular. Like any 17-year-old, she was afraid to tell her parents, so she confided in her girlfriend, Liz. Liz, being a sensible girl, persuaded Ingrid to tell her parents. You are her parent. What do you think is wrong, and how do you think it should be dealt with?

2. Julio is 62 years old, and he is worried. He has to get up two or three times in the night to urinate, and it comes so slowly! Even during the day, the stream is not as strong as it used to be. Should he be worried? What do you think may be wrong, and what should Julio do about it?

Your Turn

There is a lot of good material in this chapter for concept mapping. Start with, say, the hypophyseal portal vessels and make connections from there in chain-link fashion with the gonads, the sex organs, and relevant parts of the circulatory system and central nervous system. Use connecting words to justify the connections, like "stimulates," or "inhibits," or "facilitates," or whatever else seems appropriate. Toss a coin to determine whether you do the female or the male in your map, unless you already have a preference.

30 Human Growth and Development

Introduction

1. What are the two stages of human development?

   a. _____

   b. _____

2. What time divides the embryonic from the fetal period? _____

I. Embryonic Development (981)

1. What is a zygote? _____

2. What is cleavage? _____

3. When does the first cleavage division take place? _____

Match the following pairs.

___ 4. Fluid-filled hollow sphere   a. blastomeres
___ 5. Daughter cells   b. morula
___ 6. Mulberry-shaped group of cells   c. blastocyst
___ 7. Surrounding epithelial layer   d. inner cell mass
___ 8. Cells from which embryo will form   e. trophectoderm

9. At what point does the blastocyst shed its zona pellucida? _____

10. On the figure below, identify the following: blastocyst, trophectoderm, trophoblast cells, endometrial gland, endometrial epithelium, endometrial capillary, inner cell mass.

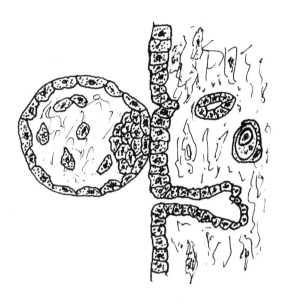

11. When does clinical pregnancy begin? _____

12. What is the function of hCG during the implantation process? _____

13. What are the two layers of the bilaminar embryonic disk?

    a. _____

    b. _____

14. At what point does cell differentiation begin to occur? _____

15. What are the three primary germ layers?

    a. _____

    b. _____

    c. _____

16. What supplies the embryo with its temporary source of nutrition prior to the third week?

    _____

17. From what cells do the extraembryonic membranes form? _____

Match the following pairs.

____ 18. Primitive respiratory and digestive structure

____ 19. Outpocket of the caudal wall of the yolk sac

____ 20. Tough, thin transparent membrane surrounding the embryo

____ 21. Protective sac around the embryo

____ 22. Precursors to spermatogonia or oogonia

a. allantois
b. chorion
c. yolk sac
d. primordial germ cells
e. amnion

23. What two tissue groups join to form the placenta?

    a. _____

    b. _____

24. What are the three main functions of the placenta?

    a. _____

    b. _____

    c. _____

25. What makes up the inner lining of the placenta? _____

26. When is the umbilical cord formed? _____

27. Why can the fetus suffer the effects of some, although not all, harmful influences in the maternal blood? _____

II. Fetal Development (989)

Match the following pairs.

___ 1. Third lunar month
___ 2. Fourth lunar month
___ 3. Fifth lunar month
___ 4. Sixth lunar month
___ 5. Seventh lunar month
___ 6. Eighth lunar month
___ 7. Ninth lunar month
___ 8. Tenth lunar month

a. nails reach tips of fingers and toes
b. the head is still dominant; lungs acquire their full shape
c. baby is at full term; fetus turns to a head-down position
d. cerebral cortex gains its typical layers; large intestine becomes recognizable
e. body is rounding; on male fetuses, testes settle into scrotum
f. fetus practices controlled breathing and swallowing
g. kidneys begin to function
h. in female fetuses, the vaginal passageway begins to develop; sweat glands appear

III. Maternal Events of Pregnancy (994)

1. List the physical indications of pregnancy at each of the following times.

    a. Three weeks _____

    b. Fifth or sixth week _____

    c. Eight weeks _____

    d. Ninth week _____

    e. Tenth week _____

    f. Twelve weeks _____

    g. Sixteen weeks _____

2. At what point does the mother begin to feel the movements of the fetus in the uterus? ___

3. When are levels of hCG the highest? _____

4. What kind of chemical is used in pregnancy testing? _____

IV. Birth and Lactation (995)

1. What does oxytocin stimulate? _____

2. What is the maximum diameter of the cervix? _____

3. What are two alternative signs indicating the onset of labor?

    a. _____

    b. _____

4. What is the cephalic position? _____

5. What feature allows the baby's head to pass through the vaginal canal? _____

6. Complete the following table.

Stage	Event(s)	Duration
First stage of labor		
Second stage of labor		
Third stage of labor		

7. What is the puerperal period? _____

8. After what day and weight are babies no longer considered premature? _____

9. What risks are inherent in late births? _____

10. What causes fraternal twins? _____

11. Which of the following do identical twins share? Which do they _not_ share?

    a. chorionic sac _____

    b. amnion _____

    c. umbilical cord _____

    d. placenta _____

12. What are conjoined twins? _____

13. What happens to the foramen ovale at birth? _____

14. What is the function of the ductus arteriosus prior to birth? _____

15. What hormone regulates milk production? _____

16. What hormone regulates milk release? _____

17. What is the function of colostrum? _____

18. What is meconium? _____

19. At what point does the mother first begin to release milk? _____

20. Describe the hormonal and ultimate physical effects of the baby's sucking action on the breast? _____

21. Can pregnancy take place during periods of lactation? _____

V. Postnatal Life Cycle (1002)

1. What are the three processes that make up human development?

   a. _____

   b. _____

   c. _____

2. During what time is a baby called a neonate? _____

3. Until what functional stage is a baby considered an infant? _____

4. Describe the neurological changes during infancy. _____

5. What years of life does childhood span? _____

6. What is adolescence? _____

7. When during adolescence does puberty take place? _____

8. What is the role of LHRH prior to puberty? _____

9. What is menarche? _____

10. Between what ages does menopause usually occur? _____

11. What are some of the results of decreased estrogen levels following menopause? _____

   _____

12. What is senescence? _____

VI. Human Inheritance and Genetic Abnormalities (1004)

1. What is the cause of Down's syndrome ? _____

2. What is the incidence of Down's syndrome? _____

3. On what three kinds of information does genetic counseling rely?

   a. _____

   b. _____

   c. _____

4. Describe the procedure of gene therapy that was followed in the case of a patient
   suffering SCID. _____

VII. When Things Go Wrong (1008)

1. What are teratogens? _____

2. How does smoking during pregnancy affect the fetus? _____

3. What are the manifestations of fetal alcohol syndrome?

   a. _____

   b. _____

   c. _____

4. What are the dangers of drug use during pregnancy? _____

5. What is the purpose of amniocentesis? _____

6. What does fetoscopy allow? _____

7. What is a benefit of chorionic villi sampling over amniocentesis? _____

8. What conditions can the alpha-fetoprotein test detect?

     a. _____

     b. _____

9. What can be done to combat hemolytic anemia? _____

10. What have recent advances in ultrasonography allowed? _____

Key Terms

adolescence 1003
adulthood 1003
blastocoele 981
blastocyst 981
blastomere 981
cleavage 981
differentiation 1002
Down's syndrome 1004
embryo 981
extraembryonic membranes 985

fetus 989
fraternal twins 999
gene therapy 1006
genetics 1004
identical twins 1000
implantation 983
inner cell mass 981
karyotype 1005
labor 995
lactation 1001

menarche 1003
menopause 1003
morula 981
parturition 995
placenta 986
puberty 1003
secondary sex characteristics 1003
senescence 1004
umbilical cord 988
zygote 981

Post Test

Matching

___ 1. The unfertilized egg

___ 2. The single-celled, fertilized egg

___ 3. A rapidly dividing, unspecialized cell migrating down the uterine (Fallopian) tube

___ 4. A hollow ball of cells

___ 5. The central, fluid-filled cavity in the hollow ball stage

___ 6. The hollow ball's outer ring of cells

___ 7. The group of cells from which the embryo proper will develop

a. blastocoel
b. inner cell mass
c. blastomere
d. blastula
e. zygote
f. trophectoderm
g. oocyte

___ 8. A protein membrane that develops from trophectoderm

___ 9. A fluid-filled sac that encloses the developing embryo

___ 10. Diminutive rootlets that penetrate the endometrium

___ 11. Has a presumed nutritive function during the second and third weeks of gestation

___ 12. Starts out as a small sac; its blood vessels become the umbilical arteries and vein

___ 13. This structure effects an intimate relationship with the endometrium and becomes the organ of exchange of gases, nutrients, hormones, and metabolites between the maternal blood vessels and those of the fetus

a. amnion
b. chorionic villi
c. chorion
d. allantois
e. yolk sac
f. placenta

___ 14. Germ layer from which most muscle
      tissue is derived
___ 15. Germ layer from which the iris
      develops
___ 16. Secretes hCG
___ 17. Source of all nervous tissue
___ 18. Epithelium of the pharynx, larynx,
      and lungs
___ 19. The thymus gland
___ 20. Its secretion inhibits menstruation
___ 21. Its secretion prevents luteal
      degeneration
___ 22. Forms a syncytium
___ 23. The original stuff of heart and
      gonads
___ 24. Forms the glandular tissues of the
      breasts
___ 25. Spleen and reticuloendothelial
      system

a. ectoderm (including neural crest)
b. endoderm
c. mesoderm
d. trophoblast
e. corpus luteum

Multiple Choice

___ 26. The principal source of estrogen and
      progesterone during pregnancy

a. blastula
b. trophoblast
c. corpus luteum
d. anterior pituitary
e. posterior pituitary

___ 27. Hairy covering of the fetus around
      the fifth month

a. caul
b. cowl
c. laguno
d. lanugo
e. lantana

___ 28. The month during which internal
      organs occupy their normal position,
      nostrils open, and thumb-sucking
      may start

a. seventh
b. sixth
c. fifth
d. fourth
e. eighth

___ 29. In males, the testes descend into the
      scrotum

a. seventh
b. sixth
c. fifth
d. fourth
e. eighth

___ 30. The chemical substance the presence
      of which is tested for in most preg-
      nancy tests

a. human chorionic gonadotropin
b. cHG
c. progesterone
d. estrogen
e. oxytocin

___ 31. An event that occurs at birth

a. the foramen ovale opens
b. the ductus arteriosus closes
c. the umbilical cord constricts
d. colostrum secretion begins
e. the lungs deflate

___ 32. A secretion rich in protein and antibodies

a. oxytocin
b. meconium
c. chorionic gonadotropin
d. prolactin
e. colostrum

___ 33. Events that occur when the infant suckles

a. GnRH inhibited
b. prolactin secreted
c. oxytocin secreted
d. both a and c
e. a, b, and c

___ 34. Signals the start of puberty

a. secretion of androgens ceases
b. secretion of LH and FSH inhibited
c. GnRH secretion stops
d. hypothalamus increases GnRH secretion
e. the amygdala inhibits the hypothalamus

___ 35. The onset of the first menstrual period

a. menopause
b. menarche
c. monarchy
d. meconium
e. PMS

Matching

___ 36. A heritable disease that used to be predominant in Ashkenazi Jews, but now is widespread

___ 37. A heritable form of muscular dystrophy

a. Down's syndrome
b. cystic fibrosis
c. Tay-Sachs
d. Duchenne's
e. Prader-Willi/Angelman syndrome

___ 38. Trisomy 21
___ 39. The most common genetic disease in the United States
___ 40. Manifestations differ, depending upon the sex of the parent from whom the genetic defect is inherited

___ 41. Transplacental infection of the fetus by this virus can cause cataracts and deafness in the child; also known as German measles

___ 42. Chemical substances that have damaging effects on developmental processes in the fetus

___ 43. This abnormality is detectable by the alpha-fetoprotein test

___ 44. Surgical delivery, bypassing the cervix and vagina

___ 45. A technique of sampling fetal cells in order to reveal the existence of any chromosomal defects by karyo-typing

a. teratogens
b. rubella
c. cesarean section
d. spina bifida
e. amniocentesis

Integrative Thinking

1. What is a "blue baby," and what are the possible causes of this condition?

2. If you were a health counselor, what advice would you give to your clents about smoking, alcohol, and mind-altering drugs?

Your Turn

This chapter marks the end of this course in anatomy and physiology. A lot of ground has been covered and, perhaps unfortunately, but of necessity, some topics have been either omitted or given in less detail than you or your instructor would have liked. Discuss the course with your study group, and please let your instructor know what you think of the coverage of the course, the textbook, and the teaching aids (such as this manual) that have been used.

ANSWERS TO POST-TEST AND INTEGRATIVE THINKING

Chapter 1

1-d; 2-b; 3-e; 4-a; 5-b; 6-b; 7-d; 8-d; 9-e; 10-c

11. Similar cells having a similar function.

12. Epithelial: covers surfaces, protective, and forms many types of glands; skin, lining of blood vessels and digestive tract. Connective: connects and supports most parts of body, main substance of bone, cartilage, tendons, skeleton, surrounding intestines, and blood vessels. Muscle: produces movement by contraction and relaxation; heart, skeletal muscles, surrounding digestive tract. Nervous: generates and transmits electrical (nervous) impulses throughout the body; brain, spinal cord, peripheral nerves.

13. See text table Body Systems.

14-b; 15.hoping you'll pick him for your answer to 14.

16-b; 17-c; 18-c; 19-a; 20-b; 21-a; 22-d; 23-a; 24-c; 25-b; 26-c; 27-a; 28-a; 29-b; 30-d.

31. A computer-guided series of X-rays for cross-sectional views of hard and soft tissues (structure)

32. Illustrates relative rate of functioning of living tissue, using uptake of radioactive metabolites (metabolic processes).

33. Video imaging of sound wave echoes from soft, moving tissues, such as fetuses. Nonradioactive.

34. Differentiates clearly between different densities of soft tissues as it measures energy emitted from hydrogen in a magnetic field. Distinguishes between diseased and healthy tissue.

Integrative Thinking

1. Make a small incision through the skin of the right thoracic wall opposite the intercostal gap closest to the position of the nodule. Probe through the incision, teasing apart underlying intercostal muscle tissue fibers, to the parietal pleural membrane. Make a small incision through the membrane directly opposite the nodule. Ascertain if the nodule is on the parietal surface of the visceral pleura. Assuming that it is, insert the thoracoscope through the opening made. Grasp the nodule with the forceps, free it from any fibrous connection, and extract it. If it is too large, use a biopsy needle to secure a small piece of the nodule and extract it.

2. This is one of the very few cases of positive feedback. It disrupts homeostasis, but with parturition (birth) accomplished, messages to the hypothalamus cease and normal mechanisms for homeostasis are quickly restored.

Chapter 2

1-e; 2-b; 3-d; 4-a; 5-f; 6-c; 7-g; 8-a; 9-b; 10-c; 11-d; 12-b; 13-e; 14-e; 15-c.

16. glucose + fructose -> sucrose + water.

$$(a) \quad + \quad (b) \quad \longrightarrow \quad (c)$$

17. power of H^+ (hydrogen ion) concentration.

18. seven (7.0), 19. Buffers are salts of weak base/strong acid or strong base/weak acid with low dissociation constant. Their action is to minimize pH changes in solution when acids or bases are added. 20. Buffers have a profound effect on homeostasis — they are the *raison d'être* of homeostasis.

21.(A) dipeptide, (B) ATP, (C) ribose, (D) steroid.

22.(a) amino group, (b) peptide bond, (c) carboxyl grp, (d) high-energy bond,(e) PO_4 group, (f) hydroxyl group.

23. building block: monosaccharides, amino acids, nucleotides;

Examples: starch, glycogen// enzymes, hair, collagen, hemoglobin// DNA, RNA.

24-a. adenine, b. ribose, (a + b = adenosine), c. 3 phosphates.

25. enzymes.

Integrative Thinking

1. You don't turn into a fish because fish protein, or that of any other organism, must be digested before being assimilated. It is broken down into its constituent amino acids, which are essentially the same in all species. From these, new proteins are synthesized in the body and become you, and not fish.

2. They have lost the ability to synthesize the digestive enzyme lactase, which digests lactose, a milk sugar. Undigested lactose is not absorbed into the blood but remains inside the intestinal tract, where it injures the lining epithelium and causes damage to varying degrees. The enzyme can be obtained at a pharmacy and taken with meals.

Chapter 3

1-e; 2-c; 3-d; 4-c; 5-a; 6-e; 7. cell adhesion molecules; 8. a progressive increase (or decrease) in concentration of a substance; 9. a molecule that facilitates transport of a substance across the cell membrane; 10. the diffusion of water through a semipermeable membrane; 11. refers to a relatively higher concentration of solute than that of a solution with which it is being compared;   12. ATP;   13. process of imbibition by means of vesicle formation by the cell membrane;   14. endoplasmic reticulum;   15. a system of microtubules and/or microfilaments that provides structural support within the cell cytoplasm.

16-d;   17-c;   18-f;   19-a;   20-e;   21-b;   22-b or c;   23-e;   24-d;   25-a;   26-f.

27a. peptide bond, 27b. Amino acid; 27c. aminoacyl-tRNA;   27d. A-site;   27e. mRNA;   28-d; 29-e;   30-h;   31-i;   32-c;   33-a;   34-b;   35-f;   36-e;   37-g;   38-d.

Integrative Thinking

1. The cells themselves, especially those of the growing skeleton, for growth hormone receptor sites. Hormones such as the growth hormone act by stimulating cells that have specific receptor sites for those hormones. If receptor sites are lacking, the hormone will have no effect. If the sites are fewer than normal, the hormone effect will be diminished correspondingly.

2. Look for mitochondria. They are probably declining in number for some reason or are lacking an enzyme or substrate essential to ATP synthesis.

3a. ribosomes and/or ER; b. mitochondria; c. microvilli and/or pinocytotic vesicles; d. lysosomes; e. lysosomes.

Chapter 4

1-a;   2-c;   3-a;   4-b;   5-c;   6-b;   7-c.

8.(a) simple squamous. (a') blood and lymphatic vessels, lining of heart, serous membranes of body cavity, etc. (b) simple cuboidal. (b') lining of many glands and their ducts, retina, kidney tubules, etc.(c) simple columnar. (c') stomach, intestine, gall bladder, digestive glands (d) stratified squamous. (d') skin, vagina, mouth, etc. (e) stratified cuboidal. (e') sweat and oil glands. (f) stratified columnar. (f') moist surfaces, e.g., larynx, soft palate, urethra, ducts of salivary and mammary glands. (g) pseudostratified columnar or transitional. (g') large excretory ducts (pseudo), urinary tract (transitional).

(For a more complete list of locations, see text.)

9-e;   10-c;   11-e;   12-a;   13-a;   14-b;   15-b;   16-e;   17-e;   18-d;   19-c;   20-b;   21-c;   22-a;   23-c;   24-d;   25-a;   26-b;   27-a;   28-c;   29-e;   30-b;   31-a;   32-d;   33-a;   34-d;   35-b;   36-c;   37-e;   38-c;   39-e;   40-a;   41-b;   42-d;   43. in order: b, a, d.

Integrative Thinking

1. Epithelium, whether thick or thin, is subject to abrasion or rapid replacement and so, if vascularized, would be subject to almost constant bleeding. It is always closely associated with a supporting layer of richly vascularized connective tissue. Cartilage is, to varying degrees, bendable and subject to compression or distortion. If there were any blood vessels in it, they would be subject to damage, resulting in extravasation or clotting. In any case, the matrix of collagen is permeable to small molecules and requires very little energy or metabolic activity.

2. Possibility (a): Well-developed tight junctions around all the endothelial cells of the blood vessels,

Possibility (b): A thickened and complete encirclement of ground substance around all the blood vessels,

Possibility (c): A combination of (a) and (b),

Possibility (d): Additional cells, such as glial cells, encasing the blood vessels.

Chapter 5

1-c; 2-b; 3-d; 4-d; 5-d; 6-b; 7-a; 8-e; 9-d; 10-c; 11-c; 12-d; 13-a; 14-e; 15-e; 16-b; 17-c; 18-c; 19-a; 20-b; 21-b; 22-b; 23-b; 24-a; 25-a; 26-c; 27-e; 28-a; 29-b; 30-d; 31-c; 32-d; 33-e; 34-a; 35-b.

36. disease, stress, malnutrition.   37. Protects scalp from sun and cold and, to some extent, bumps.   38. The half-moon, whitish area at base of the nail. Presumably white because color of blood does not penetrate this area. 39. hard keratin.   40. the remains of the eponychium. 41a. liver spots; 41b.blackheads; 41c. acne; 41d. bedsores; 41e. mole;   42-b;   43-c;   44-a; 45-e; 46-a.

Integrative Thinking

1. The nerve endings in the dermis have been destroyed, so no sensory impulses are generated. If and when the nerves regenerate, pain comes.

2. Greasy substances interfere with radiation of heat from the burn, so they aggravate the condition.

3. The germinative layer of the skin has been destroyed. Skin grafting is required.

Chapter 6

1-d; 2-c; 3-c; 4-b; 5-c; 6-a; 7-c; 8-b; 9-c; 10-d; 11-b; 12-e; 13-c; 14-d; 15-a; 16-d; 17-a; 18-b; 19-e; 20-c; 21-d; 22-d; 23-b; 24-a; 25-b; 26-c; 27-e; 28-c; 29-e; 30-a.

31. resting cartilage.   32.proliferating cartilage.   33. maturing cartilage.   34. calcifying cartilage.   35. developing trabeculae of metaphysis.

36-b; 37-d; 38-c; 39-b; 40-a; 41-d; 42-a; 43-b; 44-d; 45-d; 46-b; 47-a; 48-c; 49-d; 50-b; 51-d; 52-d; 53-a; 54-c; 55-b; 56-e; 57-b; 58-c; 59-a; 60-b; 61-d.

Integrative Thinking

1. The x-ray that showed transparent bones belonged to the elderly woman who was suffering from osteoporosis. The x-ray that showed a dark line belonged to the child. The line represented the epiphyseal growth plate. (The line was dark because x-ray photos give a negative image.)

2. Depressed thyroid function leads to decreased calcitonin levels. Along with elevated levels of PTH, bone resorption would be increased and bones would begin to demineralize.

3. It would impair his ability to fight infections, because the white cells involved in immunity would not be produced in his bone marrow, where these white cells are formed.

Chapter 7

1-c; 2-b; 3-e; 4-d; 5-a; 6-f; 7-c; 8-b; 9-g; 10-b; 11-a; 12-d; 13-h; 14-e; 15-f; 16-f; 17-c; 18-g; 19-a; 20-d; 21-h; 22-f; 23-b; 24-c; 25-e; 26-e; 27-a; 28-b; 29-d; 30-f; 31-b; 32-d; 33-c; 34-d; 35-a; 36-e; 37-a; 38-d; 39-g; 40-b; 41-h; 42-f; 43-c; 44-f; 45-a; 46-c; 47-f; 48-d; 49-c; 50-e; 51-d; 52-b; 53-d; 54-e; 55-f; 56-g; 57-c; 58-d; 59-a; 60-b; 61-c; 62-b; 63-f; 64-a; 65-g; 66-d; 67-e; 68-h; 69-f; 70-g; 71-b; 72-c; 73-a; 74-i; 75-g; 76-e; 77-d; 78-**7**; 79-**12**; 80-**5**; 81-**5**; 82-f; 83-g; 84-b; 85-a; 86-h; 87-e; 88-d; 89-c; 90-f; 91-e; 92-c; 93-a; 94-d; 95-f.

Integrative Thinking

1. Maxilla, mandible

2. From the nasal cavity superiorly through the sphenoidal bone, through the sphenoidal sinus, through the superior wall of the sphenoidal sinus, and into the sella turcica.

3. Whiplash can be mild, serious, or very serious, depending on the extent of injury to cervical vertebrae, muscles, ligaments, or spinal cord. The mildest injury involves muscle and ligament strain, which, through painful for a considerable length of time, will repair itself completely. Dislocation of cervical vertebrae and/or tearing of ligaments can be more serious, and may result in permanent "pain-in-the-neck" discomfort and stiffness. Damage to the spinal cord is the most serious consequence and may result in permanent paralysis distal to the site of damage.

4. By lumbar puncture (spinal tap) with a needle and syringe in the midline between the third and fourth or fourth and fifth lumbar vertebrae into the spinal canal of the spinal cord.

Chapter 8

1-k; 2-e; 3-c; 4-l; 5-j; 6-a; 7-b; 8-m; 9-g; 10-h; 11-d; 12-d; 13-b; 14-a; 15-i; 16-f; 17-a; 18-b; 19-i; 20-b; 21-b; 22-f; 23-b; 24-a; 25-e; 26-d; 27-c; 28-c; 29-f; 30-d.

31. os coxa (hip bone). 32. head of femur. 33. greater trochanter. 34. medial condyle. 35. tibia. 36. metatarsals;

37. cuboid. 38. navicular. 39. talus. 40. calcaneus.

Integrative Thinking

1. The clavicle is a thin-walled bone. It is in a very vulnerable position, and is the only direct skeletal connection between the axial skeleton and the rest of the pectoral limb.

2. The bones of elderly people tend to be much thinner than those of younger persons. Men's bones are generally thicker than women's at all stages of adult life, despite the thinning that occurs in old age. Most elderly women have osteoporosis as a consequence of the dramatic change in sex hormones that begins following the onset of menopause. Exogenous estrogen and the control of calcium and phosphorus metabolism are preventives for the loss of bone tissue that otherwise would occur during these years.

3. The scaphoid. The reflex tendency is to abduct (turn outward) the hand at the wrist, which places the scaphoid in direct line with the radius of the forearm and with the trapezium and

trapezoid of the wrist and the first and second fingers. The force of the fall is thus concentrated along this line, which, on impact, pushes the hand upward at a sharp angle.

Chapter 9

1-a; 2-e; 3-c; 4-b; 5-d; 6-a; 7-e; 8-b; 9-c; 10-f; 11-d; 12-e; 13-a; 14-d; 15-b; 16-e; 17-f; 18-c; 19-a; 20-d; 21-b; 22-a; 23-d; 24-b; 25-a; 26-c; 27-c; 28-c; 29-b; 30-e; 31-c; 32-b; 33-f; 34-a; 35-b; 36-b; 37-h; 38-d; 39-e; 40-g; 41-d; 42-a; 43-c; 44-e; 45-b.

Integrative Thinking

1. A baseball may not weigh very much (a little over 5 ounces), but the force required to accelerate the flight of a ball from zero to speeds of 85 to 95 miles per hour in a matter of milliseconds is obviously considerable. But this is only part of the problem. In order to attain this speed, the entire body must contribute to this force — the legs, the torso, and both arms. The joints most affected are the elbow, the shoulder, and the wrist of the throwing arm. The elbow receives the most punishment, because of the complexity of its motion. It is a hinge joint with two sets of articulations but only one joint capsule plus a large bursa for lubricaton. There is a radial collateral ligament that reinforces the joint laterally, and an ulnar collateral ligament medially. Several muscles have either origins or insertions near the joint, creating a highly complex system of tendons and ligaments in the area. The rotator cuff of the shoulder joint, although less severely affected by the throwing action, is often affected nevertheless, as are the muscles involved in this joint. (They are considered in Chapter 11.) One can readily appreciate the complexity of the maneuver in this way: With the elbow bent, twist the arm backward as far as it will go at ear level, hyperextend the hand at the wrist, fingers flexed, grasping an imaginary ball tightly, and then throw as hard and as fast as possible! Warning: Go through the motions a bit slowly a few times at first, or a single throw will make your arm quite sore. The pitcher must allow a couple of days, at least, to avoid more or less permanent damage to the elbow joint.

2. With time, the intervertebral disks become compressed, compacted, and more fibrous. The compacting accounts for most of the loss of height, but some results from the increased curvature of the vertebral column in the upper thoracic region that is cause by contraction of the fibrous components of the disks.

3. The ligaments maintain the integrity of the joint, holding it together against the muscular forces that tend to displace or dislocate it.

4. The TMJ syndrome is a multiplicity of symptoms attributable to malalignment of the temporomandibular joint on one or both sides. These symptoms include pain, which may be moderate to severe, intermittent or continuous, in the neck and/or shoulders, and accompanied in some cases by disorientation, vertigo, and ringing in the ears (tinnitus).

Chapter 10

1-d; 2-b; 3-a; 4-a; 5-c; 6-a; 7-c; 8-b; 9-a; 10-a; 11-b; 12-b; 13-e; 14-b; 15-a; 16-d; 17-e; 18-c; 19-b; 20-a; 21-b; 22-e; 23-d; 24-a; 25-c; 26-f; 27-c; 28-a; 29-g; 30-e; 31-f; 32-d; 33-h; 34-f; 35-e; 36-d; 37-b; 38-d; 39-c; 40-e; 41-e; 42-a; 43-d; 44-a; 45-b; 46-d; 47-d; 48-e; 49-a; 50-c; 51-b; 52-b; 53-e; 54-d; 55-c;

56-a; 57-acb; 58-abc; 59-cab; 60-caa; 61-bab; 62-abc; 63-abc; 64-bac; 65-b; 66-a; 67-b; 68-d; 69-b; 70-a; 71-a; 72-c; 73-c.

Integrative Thinking

1. The skeletal muscles become paralyzed and slowly undergo atrophy. If the muscles of the thoracic cage become paralyzed, death quickly ensues from asphyxiation. If caught in time, the patient can be placed in a device that will cause air to move in and out of the lungs. The first such device was the iron lung, an airtight chamber that was attached to an air compressor that would alternately raise and lower the pressure in the chamber. The patient would lie in the chamber, with his or her head outside. When the chamber pressure was below that of the surrounding air, room air would be drawn into the lungs. When the chamber pressure exceeded that of the room, air would be forced out of the lungs. Smaller and much lighter chambers that can be worn like a jacket or life preserver succeeded the old iron lung and have the great advantage of being portable. The wasting effect of muscle atrophy in the limbs can be slowed considerably by physical therapy, making extensive use of massage and whirlpool baths.

2. The arrangement of the heart muscles is such that, when they contract during ventricular systole, the heart is subjected to a sort of wringing action, which empties the chambers more thoroughly than straight compression would do. Moreover, with the end-to-end arrangement of muscle cells, which pass on the stimulus from cell to cell, the heart contracts more gradually, as in a progressive wave. Thus, the action is gentler on the delicate valves that are forced open and shut by the pressure changes of systole and diastole.

Chapter 11

1-e; 2-c; 3-a; 4-j; 5-f; 6-d; 7-h; 8-g; 9-b; 10-i; 11-h; 12-d; 13-b; 14-a; 15-g; 16-e; 17-c; 18-f; 19-b; 20-c; 21-c; 22-a; 23-b; 24-a; 25-c; 26-a; 27-a; 28-b; 29-d; 30-d; 31-e; 32-c; 33-b; 34-a; 35-c; 36-d; 37-b; 38-e; 39-b; 40-a.

Integrative Thinking

1. The sartorius crosses over from a lateral origin to a medial insertion, which accounts for part of the difference. More significantly, its insertion is distal to the knee joint, whereas the distal tendon of the rectus femoris, although functionally inserting at about the same distance distally as the tendon of the sartorius, actually is interrupted by inserting on the patella, which, in turn, inserts on the tibia beside that of the sartorius.

2. In the olden days, tailors traditionally sat cross-legged on a bench or tabletop. The sartorius made this position possible by rotating the leg medially. Try it!

3. Cutting these tendons crippled the animals so that they could not even stand up, and thus indirectly crippled the cavalrymen.

4. In a leverage system, the fulcrum does not move in either direction.

Chapter 12

1-d; 2-e; 3-a; 4-b; 5-d.

6. nodes of Ranvier. 7. Schwann; 8. oligodendrocytes. 9. ganglion. 10. satellite cells.

404

11-c; 12-d; 13-b; 14-e; 15-a; 16-i; 17-c; 18-h; 19-a; 20-d; 21-g; 22-j; 23-f; 24-b; 25-e.

26. sodium-potassium pump. 27. sodium and potassium. 28. negative. 29. resting membrane potential. 30. ATP.

31-d; 32-a; 33-d; 34-c; 35-b; 36-d; 37-c; 38-a; 39-a; 40-b; 41-f; 42-g; 43-e; 44-f; 45-d; 46-b; 47-c; 48-a; 49-e; 50-d; 51-d; 52-f; 53-c; 54-a; 55-d; 56-b; 57-c; 58-a; 59-g; 60-c; 61-f; 62-b; 63-i.

Integrative Thinking

1. Brain tumors invariably are of the nonneuronal cells – the neuroglia.

2. The resting membrane potential would be adversely affected. Due to a lowered concentration of potassium in the body fluids, the Na$^+$/K$^+$ ratio would be shifted, overpowering the self-regulatory ability of the sodium-potassium pump. The result would be a hyperpolarization of the cell membrane. This condition is believed to contribute to the muscle weakness that characterizes potassium deficiency.

3. Curare is used to prevent muscle spasms that sometimes occur during surgery and interfere with delicate surgical procedures.

Chapter 13

1-b; 2-d; 3-a; 4-c; 5-d; 6-a; 7-d; 8-c; 9-a; 10-d; 11-c; 12-c; 13-a; 14-b; 15-d; 16-c; 17-a; 18-a; 19-d; 20-c; 21-a; 22-c; 23-b; 24-d; 25-a; 26-d; 27-d; 28-a; 29-c; 30-b; 31-g; 32-d; 33-i; 34-b; 35-a; 36-h; 37-e; 38-c; 39-j; 40-f; 41-c; 42-a; 43-b; 44-e; 45-e; 46-c; 47-a; 48-e; 49-c; 50-c; 51-e; 52-a; 53-d; 54-b; 55-b; 56-d; 57-a; 58-c; 59-b; 60-a; 61-e; 62-c; 63-a; 64-b; 65-d; 66-c; 67-b; 68-b; 69-d; 70- ; 71-b; 72-d; 73-a; 74-c; 75-a.

Integrative Thinking

1. Margaret is either popping amphetamines or getting hooked on cocaine. She needs professional help, and somebody should be alerted to the high probability that her new "friends" should be looked into.

2. Two possibilities: (a) The right side of the cerebellum has been damaged, or (b), less likely, the primary somesthetic association area of the left parietal lobe of the cerebrum.

3. It is now thought that the somesthetic association area of the parietal lobe of the cerebrum is generating impulses that are passed on to the limbic system, where they are given emotional content (i.e., in this case, unpleasant sensations).

Chapter 14

1-a; 2-d; 3-e; 4-d; 5-b; 6-h; 7-j; 8-d; 9-l; 10-a; 11-g; 12-i; 13-k; 14-c; 15-f; 16-b; 17-m; 18-e; 19-d; 20-c; 21-f; 22-b; 23-a; 24-e; 25-c; 26-a; 27-b; 28-a; 29-b; 30-b; 31-d; 32-a; 33-c; 34-e; 35-b; 36-c; 37-c; 38-a; 39-b; 40-d.

Integrative Thinking

1. (a) Spinal meningitis. (b) Biochemical analysis of a spinal tap. (c) The presence of pus in the spinal tap would indicate a bacterial infection of the meninges. If no pus, the causative agent is probably a virus.

2. So as not to impede the action of the agonists. The crossed extensor reflex serves to push the body away from the cause of the reflex, thereby supporting the withdrawal action.

3. Proprioception, touch pressure, and two-point discrimination would be affected on the side contralateral to the lesion at all levels inferior to T4 (cf. text table 14.1).

Chapter 15

1-c; 2-f; 3-g; 4-d; 5-g; 6-b; 7-e; 8-a; 9-e; 10-d; 11-a; 12-h; 13-d; 14-c; 15-j; 16-f; 17-g; 18-k; 19-l; 20-e; 21-i; 22-b; 23-i; 24-k; 25-l; 26-g; 27-l; 28-i; 29-l; 30-k; 31-j; 32-i; 33-h; 34-g; 35-a; 36-b; 37-c; 38-e; 39-f; 40-d; 41-b; 42-e; 43-a; 44-c; 45-d; 46-c; 47-a; 48-d; 49-e; 50-b.

Integrative Thinking

1. (a) CN V (trigeminal)

   (b) CN VIII (vestibulocochlear)

   (c) CN XII (hypoglossal)

   (d) CN III (oculomotor)

2. When fitted, the knot of the noose rests against the base of the skull so that when the body falls, the head is violently rotated forward and downward. This causes the vertebral column to bend at the atlas-axis articulation, with the result that the odontoid process is forced into the spinal canal, crushing the spinal cord at this point. The shock instantly renders the victim unconscious, and the motor tracts are destroyed, causing cessation of respiration, among other functions. In effect, the victim dies quickly of asphyxiation.

Chapter 16

1-b; 2-e; 3-a; 4-a; 5-c; 6-c; 7-e; 8-a; 9-b; 10-d; 11-g; 12-d; 13-c; 14-j; 15-h; 16-i; 17-a; 18-e; 19-f; 20-b; 21-b; 22-a; 23-b; 24-a; 25-d; 26-a; 27-c; 28-b; 29-a; 30-b; 31-e; 32-b; 33-a; 34-a; 35-c; 36-b; 37-e; 38-a; 39-c; 40-d.

Integrative Thinking

1. Typical visceral reflexes, such as sweating, peristalsis, and vasomotion, are transmitted and arc no higher in the central nervous system than the medulla oblongata and so do not reach cerebral levels, where they must in order for one to be conscious or aware of them. In fact, it is now thought that many visceral reflexes can be assigned to the enteric division of the autonomic nervous system, which may have no direct connection to the central nervous system at all.

2. The "seat" of the emotions, so to speak, is the hypothalamus, which has direct synaptic connection with the autonomic nervous system via the brainstem, where most regulatory centers of the autonomic nervous system are located.

Chapter 17

1-a; 2-d; 3-b; 4-b, a, d, e, c; 5-e; 6-c; 7-e; 8-a; 9-c; 10-d; 11-b; 12-c; 13-b; 14-f; 15-c; 16-d; 17-a; 18-b; 19-c; 20-e; 21-f; 22-g; 23-d; 24-e; 25-h; 26-c; 27-a; 28-f; 29-b; 30-b; 31-f; 32-g; 33-a; 34-e; 35-c; 36-d; 37-e; 38-c; 39-c; 40-a; 41-b; 42-b; 43-c; 44-a; 45-b; 46-c; 47-b; 48-b; 49-a; 50-c; 51-d; 52-b; 53-f; 54-a; 55-e; 56-c; 57-b; 58-e; 59-a; 60-d; 61-b; 62-e; 63-e; 64-c.

Integrative Thinking

1. The flashes appear to be in the opposite side of the eye. This is because the brain is constructed to compensate for the "camera effect"— that is, the light rays cross as they pass through the lens, so that left becomes right and right becomes left. The brain compensates for this, reading right for left and left for right. When you rub the nasal side, you stimulate the rods in that side of the retina to fire, but your brain reads the effect the opposite way.

2. This test determines the internal pressure of the eyeball. Glaucoma is caused by a buildup of intraocular pressure, usually owing to blockage of the canal of Schlemm, which drains the anterior chamber as the pressure rises above the optimum value. Any value above this optimum implies the development of glaucoma, which inevitably leads to blindness.

3. Macula degeneration. In the elderly there is a tendency for the tiny blood vessels behind the retinal layer in the macula area to leak, creating small clots that interfere with nourishment of the retinal cells. As a consequence, these cells die. There is no cure, but the progress of macula degeneration can sometimes be slowed by laser therapy.

4. The answer appears to be that in this case deafness is due to calcification between the otic ossicles, preventing their movement in response to sound waves. The internal ear appears to be functional because it is receiving sound vibrations through the skull. Hearing aids that compensate for this condition are available, and in some cases the ossicles can be unfused and restored to normal function.

Chapter 18

1-f; 2-b; 3-b; 4-e; 5-g; 6-e; 7-a; 8-c; 9-e; 10-a; 11-f; 12-g; 13-b; 14-b; 15-d; 16-b; 17-d; 18-a; 19-c; 20-f; 21-e; 22-c; 23-f; 24-a; 25-c; 26-c; 27-b; 28-b; 29-d; 30-c; 31. a>d>c>f>b>g>e>h>i; 32-d; 33-b; 34-a; 35-d; 36-a; 37-c; 38-d; 39-d; 40-a; 41-a; 42-d; 43-e; 44-c; 45-f; 46-b; 47-g; 48-c; 49-b; 50-d.

Integrative Thinking

1. Perhaps the most likely explanation is that there are insufficient of receptors for growth hormone on the cells that should have them. A parallel example of this situation appears to apply to Type II diabetes, which figures in the problem below. Since there is no lack of GH in these people, the hypothalamus must be secreting growth hormone releasing hormone, and growth hormone inhibiting hormone must not be a problem either. There is a remote probability of an imbalance in thyroid secretions, for they, too, are involved in the metabolism of all tissues involved in growth.

2. The tests for steroids are more likely to show by-products of steroid metabolism than the steroids themselves. Hence, the presence of unmetabolized steroids in any excessive amount would be expected if they had been added directly to the urine itself. Consequently, Jan's

claim may have some validity. On the other hand, there might be indications of prior or habitual use of steroids in venues other than urine. These indications differ somewhat between males and females, as discussed in the textbook. The presence of such indications would certainly prejudice the judgment of the examiners and lead to disqualification.

3. Melanie has Type I diabetes, which is insulin-dependent, whereas Jacob has Type II diabetes, which is insulin-independent.

Chapter 19

1-e; 2-d; 3-a; 4-d; 5-a; 6-b; 7-b; 8-a; 9-d; 10-b; 11-c; 12-b; 13-c; 14-d; 15-a; 16-c; 17-b; 18-d; 19-a; 20-c; 21-g; 22-c; 23-e; 24-f; 25-d; 26-h; 27-b; 28-a; 29-b; 30-d; 31-c; 32-e; 33-a; 34-f; 35-d; 36-a; 37-e; 38-b; 39-c; 40-c; 41-a; 42-f; 43-e; 44-b; 45-d; 46-b; 47-a; 48-d; 49-c; 50-d; 51-d; 52-e; 53-d; 54-b; 55-c; 56-a; 57. b>c>a>d.

Integrative Thinking

1. Immediate problem: shortness of breath following even very slight physical exertion. Hematocrit or hemocytometer would show a lower percent hemoglobin and r.b.c. count than that of the indigenous population. Hypoxia is induced by low partial pressure of oxygen in atmosphere at high altitude. This would induce erythropoietin synthesis and secretion by the kidneys, which would stimulate the myelloid tissue to produce more erythrocytes. After a few days, increased r.b.c. count and hemoglobin level would render the person adapted to the high altitude, and symptoms would either disappear or be less severe. Most persons in good health can adapt to altitudes

up to approximately 15,000 feet.

2. The baby could have viral hepatitis, but more likely it has what is known as hemolytic disease of the newborn. If the latter, one would expect that (a) the mother is Rh negative, (b) the father is Rh positive, (c) the infant is Rh positive, and (d) the mother had a previous pregnancy with an Rh positive child.

3. (a) Identify the situation: names of the affected persons (with permission) and neighborhood. (b) Identify the offending party — the factory, its products or procedures, its record of infractions. (c) Describe the two major kinds of leukemias, acute and chronic: the man probably has one type and the little girl the other, how they may be caused by environmental pollutants, either physical (radiation) or chemical, the normal course of the disease, therapeutic procedures and prognosis. Probability that the only likely connection between the two victims is the pollution, otherwise coincidence. (d) Identify the agencies or action groups or individuals that are doing something about the present situation.

4. As described in the textbook, bone marrow replacement is hazardous because it requires that the patient be exposed to high levels of ionizing radiation and/or chemotherapy in order to kill off all the cells in the bone marrow, which includes the cells of the immune system. This renders the patient vulnerable to infection until reseeding of the marrow with healthy, donor cells is established.

Chapter 20

1-c; 2-a; 3-d; 4-b; 5-c; 6-d; 7-a; 8-d; 9-b; 10-d; 11-a; 12-e; 13-e; 14-a; 15-f; 16-c; 17-d; 18-a; 19-e, f; 20-a; 21-b; 22-b; 23-c; 24-a; 25-a; 26-d; 27-e; 28-e; 29-a; 30-d; 31-b; 32-c; 33-c; 34-e; 35-a; 36-b; 37-d; 38-e; 39-c; 40-a; 41-b; 42-c; 43-d; 44-d; 45-a; 46-b; 47-e; 48-c; 49-f; 50-d; 51-c; 52-e; 53-b; 54-d; 55-a; 56-a; 57-e; 58-c; 59-d; 60-b.

Integrative Thinking

1. Three diagnostic techniques: echocardiography, nuclear scanning, dynamic spatial reconstruction.

Two therapies: balloon angioplasty, infusion of so-called "clot-buster" enzymes, such as streptokinase and recombinant tissue plasminogen activator (rt-PA).

2. AV block.

3. If so placed, electrical stimulating signals would be transmitted to all four chambers simultaneously, and they would all contract at the same time. Result: a knot!

Chapter 21

1-a; 2-d; 3-a; 4-e; 5-b; 6-b; 7-c; 8-d; 9-a; 10-c; 11-b; 12-e; 13-c; 14-b; 15-d; 16-a; 17-e; 18-d; 19-e; 20-a; 21-d; 22-c; 23-c; 24-b; 25-a; 26-f; 27-i; 28-c; 29-h; 30-e; 31-d; 32-b; 33-g; 34-d; 35-e; 36-b; 37-c; 38-a; 39-c; 40-b; 41-d; 42-a; 43-b; 44-d; 45-c; 46-a; 47-e; 48-e; 49-b; 50-e; 51. (in any order) arterial pressure, right atrial pressure, resistance, venous pump, hydrostatic pressure.

Integrative Thinking

1. All the substances that skeletal muscle needs from the blood, such as oxygen and glucose, are small and readily diffusible through the capillary walls; similarly the wastes

from skeletal muscle metabolism are simple molecules and/or are readily diffusible.

2. The initial response to exposure to cold is the dilation of capillaries in the skin, warming the skin and causing it to redden, which does not show if the skin is pigmented. Continued exposure to cold results in sufficient lowering of body temperature to cause these capillaries to constrict in order to conserve heat, resulting in blanching of the skin. Finally, the shivering reflex follows, which generates body heat.

3. The first cause of these symptoms that comes to mind is hypertension, because this is a common cause of the symptoms described, especially in the over-50 age group, as well as in heavy smokers. Although the condition is not necessarily inherited, the propensity for it may be. Other causes include diabetes, obesity, stress, and to a less extent, certain metabolic abnormalities.

Chapter 22

1-c; 2-a; 3-e; 4-d; 5-b; 6-c; 7-a; 8-b; 9-d; 10-e; 11-b; 12-e; 13-d; 14-d; 15-a; 16-c; 17-b; 18-b; 19-e; 20-c; 21-c; 22-c; 23-a; 24-b; 25-d.

Integrative Thinking

1. The introduction of foreign substances, either antigenic or irritative, elicits hypertrophy. Mitosis in the germinal centers increases, producing more lymphocytes, and antibody activity goes into high gear. Edema might result upstream of the node as a result of pressure backup.

2. Lymphatic drainage of the head, jaw, and neck is filtered by lymph nodes in the throat region. Swollen nodes imply the presence of an infection in the area drained by the nodes, so look there.

3. If there is lymph node swelling in the throat or neck area, the cause of the toothache is most likely an infection of the tooth and/or the surrounding tissues. Swelling of the soft tissues in the general area indicates a backup of lymph, as noted in question 1, and can be correlated with the severity of the infection.

4. To look for the presence of metastasis. Neighboring lymph nodes are the first place to look for metastatic growth of a neoplasm. If any is present, the node would be excised with the hope that, together with removal of the primary site, the whole tumor will be removed and a cure effected.

Chapter 23

1-d; 2-b; 3-a; 4-a; 5-b; 6-e; 7-b; 8-d; 9-a; 10-c; 11-d; 12-c; 13-a; 14-e; 15-b; 16-g; 17-f; 18-d; 19-i; 20-b; 21-a; 22-j; 23-k; 24-c; 25-e; 26-d; 27-b; 28-e; 29-d; 30-e; 31-c; 32-a; 33-d; 34-c; 35-e; 36. d>f>c>a>b>e; 37-e; 38-c; 39-a; 40-g; 41-d; 42-f; 43-b; 44-c; 45-d; 46-e; 47-a; 48-c; 49-d; 50-d; 51-b; 52-e; 53-a; 54-f; 55-a; 56-a; 57-b; 58-c; 59-d; 60-b.

61. any four of following: HIV-1 Eliza, Western Blot, HIV-1/HIV-2 blood test, 10-minute blood test, PCR blood test, urine test.

Integrative Thinking

1. Complement fixes onto antigens, making them susceptible to antibody, as well as promoting phagocytosis and the inflammatory response. All these actions result in rapid destruction of antigen cells and rejection of organ transplants from another animal species (xenotransplants). If complement is inhibited, critical aspects of the immune process will be rendered inoperative.

2. Either a continuous infection involving some unidentified antigen or the development of uncontrolled division of one of the stem cells (myeloma) in the hemopoietic tissue.

3. (a) Allergic reaction; (b) IgE.

Chapter 24

1-a; 2-e; 3-c; 4-c; 5-b; 6-d; 7-a; 8-d; 9-e; 10-b; 11-e; 12-b; 13-d; 14-a; 15-c; 16-h; 17-e; 18-f; 19-g; 20-c; 21-a; 22-d; 23-b; 24-a; 25-h; 26-c; 27-f; 28-d; 29-g; 30-b; 31-e; 32-c; 33-e; 34-a; 35-b; 36-f; 37-d.

38. pulmonary. 39. bronchial.

40-b; 41-d; 42-c; 4-a; 44. b, b, a; 45. a, a, a; 46. c, c, a; 47. a, a, c; 48. c, c, a; 49-e; 50-b; 51-c; 52-d; 53-a; 54-a; 55-e; 56-c; 57-b; 58-c; 59-f; 60-a; 61-d; 62-e; 63-b; 64-e; 65-e.

Integrative Thinking

1. Emphysema, coronary disease, asthma, reduced partial pressure of oxygen in the atmosphere, lack of physical fitness, obesity, infectious disease of the respiratory system, such as pneumonia, tuberculosis, or cystic fibrosis.

2. The partial pressure of oxygen, as with any gas, decreases with altitude. Pressurizing the cabin results in simulating lower altitudes, and the partial pressure of oxygen is maintained at a safe level.

3. Divers substituted helium for nitrogen in order to avoid getting the "bends" when returning from a dive. Rapid decrease in pressure causes nitrogen, dissolved in the blood at high pressure, to come out of solution and to form bubbles in the alveolar tissues, which is life-threatening. In addition, because nitrogen is fat soluble, it tends to accumulate in fatty tissues and certain joints. Bubbles forming in these tissues can be painful to the point of inducing shock.

Chapter 25

1-c; 2-d; 3-a; 4-g; 5-b; 6-f; 7-e; 8-g; 9-e; 10-a; 11-d; 12-f; 13-b; 14-c; 15-a.

16. two. 17. canines. 18. incisors. 19. periodontal ligament. 20. *32.*

21-d; 22-f; 23. c; 24-a; 25-e; 26-g; 27-b; 28-f; 29-g; 30-c; 31-e; 32-h; 33-d; 34-a.

35. tongue. 36. saliva. 37. peristalsis. 38. greater omentum. 39. pyloric sphincter. 40. rugae.

41-b; 42-d; 43-a; 44-c; 45-a; 46-d; 47-a; 48-d; 49-c; 50-e; 51-b; 52-a; 53-d; 54-c; 55-e; 56-g; 57-f; 58-h; 59-b; 60-c; 61-g.

62. (any six in any order) proteins, carbohydrates, lipids, nucleic acids, water, vitamins, ions, trace elements.

63.

CCK	a	f	k, j, m
GIP	a	e	m
Gastrin	c	g, h	i
Secretin	a	e	j, m, n

64. hepatic artery, hepatic vein, hepatic portal vein (in any order).

65-c; 66-d; 67-a; 68-a; 69-b; 70-b; 71-d; 72-c; 73-f; 74-e; 75-a.

Integrative Thinking

1. Food should be taken slowly and in small amounts. Proteins, such as meat, eggs, and nuts, should be finely minced and incorporated in a watery broth; other foods should also be well chewed and quite soupy. Normally the stomach completes the jobs of mastication, grinding, and making the food soupy before it injects it in measured amounts through the pyloric sphincter into the duodenum. Moreover, proteins are partly hydrolyzed and thoroughly minced. Milk should bebrought to the boiling point and then cooled before drinking, or taken in the form of yogurt or soft cheese. Carbohydrates should be well cooked and moist or liquefied. Fats should be ingested sparingly, because biliary flow may be reduced.Most importantly, one should eat very small amounts at a sitting, and the sittings should be separated by intervals of a half hour or so, because digestion in the small intestine is a slow process. Finally, food should be aseptic -that is, uncontaminated by microorganisms, such as you might encounter in fresh fruit or in any raw food, for that matter. The intact stomach is a great sterilizer, but the intestines are not.

2. The baby probably has pyloric stenosis, or extreme narrowing, if not complete closure, of the pyloric valve, which prevents food from moving from the stomach into the small intestine. This condition can be corrected surgically.

3. Atrophy of skeletal muscle, severe constipation, dangerously lowered blood pressure, increased susceptibility to infection, dehydration, and electrolyte imbalance. In prolonged cases, a point of no return may be reached, and untimely death soon follows.

Chapter 26

1-d; 2-e; 3-a; 4-f; 5-c; 6-b; 7-g; 8-b; 9-c; 10-e; 11-e; 12-c; 13-b; 14-a; 15-b; 16-b; 17-d; 18-a; 19-c; 20-e; 21-a; 22-d; 23-b; 24-c; 25-e; 26-b; 27-e; 28-d; 29-a; 30-c; 31-c; 32-d; 33-a; 34-d; 35-b; 36-c; 37-a; 38-d; 39-e; 40-b; 41-c; 42-a; 43-e; 44-c; 45-e; 46. (in any order) calcium, chlorine, magnesium, phosphorus, potassium, sodium, sulfur. 47. (any seven), cobalt, copper, fluorine, iodine, manganese, molybdenum, selenium, zinc.

48-d; 49-a.

Integrative Thinking

1. Scurvy is characterized by anemia, low resistance to infection, swollen and bleeding gums, bruises, and prostration. Cause: dietary — lack of fresh vegetables and fruit. Remedy: fresh fruit, such as oranges and lemons, fresh leafy vegetables, and potatoes; sources of vitamin C.

2. Consider vitamins B_5, biotin, and inositol, singly or in combination.

3. Vitamin B_2.

Chapter 27

1-e; 2-d; 3-c; 4-e; 5-a; 6-c; 7-f>j>b>h>e>l>d>a>i>c>k>g.

8. a>e>h>c>g>b>f>d. 9. the nephron (loop of Henle).

10-c; 11-a; 12-e; 13-b; 14-d; 15-b; 16-g; 17-d; 18-c; 19-h; 20-a; 21-b; 22-i; 23-f; 24-e; 25-i; 26-c; 27-d; 28-a; 29-b; 30-g; 31-b; 32-e; 33-d; 34-a; 35-c; 36-b; 37-f; 38-c; 39-a; 40-d; 41-e.

Integrative Thinking

1. A variety of certain unrelated pollutants, including insecticides, which farmers use extensively, and certain hydrocarbons, such as benzene, carbon tetrachloride, toluene, and mercuric compounds, which are components of some paints and solvents to which both house painters and autobody repair workers are chronically exposed. If the correlation between exposure to these pollutants and the incidence of acute kidney failure should be proved to be statistically significant when compared to populations not exposed to them, then acute renal failure can be considered an occupational disease.

2. Pyelonephritis.

3. The nephrons would probably resemble those of certain Earthling marine fishes and desert mammals, which have a very small glomerulus and glomerular capsule, which severely limits the filtration rate, and well developed tubules — especially the nephron loops and collecting ducts — for efficient water reabsorption and electrolyte control.

Chapter 28

1-b; 2-e; 3-d; 4-e; 5-b; 6-c; 7-b; 8-d; 9-a; 10-d; 11-b; 12-f; 13-c; 14-d; 15-a; 16-e.

17. (in any order) a. bicarbonate buffers, b. phosphate buffers, c. protein buffers. 18. protein. 19. resist changes in pH. 20. protein. 21. by controlling rate and depth of respiration.

22-b; 23-e; 24-a; 25-e; 26-b; 27-a; 28-c; 29-e; 30-d.

Integrative Thinking

1. These are the main symptoms of hypernatremia. Get drinking water for her as quickly as possible, and in the meantime take the chips away from her – they are loaded with salt. Make her rest while you're after the water. She could be in trouble if you don't hurry, for this condition can be quite serious. Finally, make sure you drink some water, too.

2. Edema may have several causes: (a) plasma protein may leak out of the small blood vessels if they have been damaged by an injury; (b) hepatitis may lead to generalized edema by interstitial accumulation of plasma water; (c) high blood pressure may force water by hydrostatic pressure out of the capillaries and into the interstitial compartment. Other pathological conditions may have similar consequences.

Chapter 29

1-b; 2-e; 3-f; 4-c; 5-g; 6-d; 7-a; 8-d; 9-e; 10-a; 11-c; 12-b; 13-d; 14-b; 15-c; 16-i; 17-j; 18-h; 19-a; 20-g; 21-e; 22-f; 23-c; 24-a; 25-e; 26-b; 27-f; 28-a; 29-d; 30-b; 31-a; 32-a; 33-g; 34-b; 35-e; 36-f; 37-d; 38-h; 39-c; 40-f.

41. estrogen and progesterone. 42. oogenesis. 43. corpus luteum. 44. corpus albicans. 45. stratum functionali.

46-d; 47-e; 48-b; 49-a; 50-c; 51-e; 52-d; 53-b; 54-a; 55-b.

Integrative Thinking

1. Ingrid may have an ovarian cyst. The condition may clear up by itself, but in any case, she should see a gynecologist. There is always the possibility of an ovarian tumor, so Ingrid should get right on it. As her parent, I would discuss the problem with sympathy and understanding, and reassure Ingrid that her very best friends are her mother and father.

2. Julio need not be worried, but he should be concerned. He has prostatitis, or swelling of the prostate gland. The majority of men over age 60 or so have this complaint. Julio should see a physician, preferably a urologist, and be examined to see if he has a cancerous or precancerous condition. Both prostatitis and cancer of the prostate are treatable.

Chapter 30

1-g; 2-e; 3-c; 4-d; 5-a; 6-f; 7-b; 8-c; 9-a; 10-b; 11-e; 12-d; 13-f; 14-c; 15-a; 16-d; 17-a; 18-b; 19-b; 20-e; 21-d; 22-d; 23-c; 24-a; 25-c; 26-c; 27-d; 28-b; 29-e; 30-a; 31-b; 32-e; 33-e; 34-d; 35-b; 36-c; 37-d; 38-a; 39-b; 40-e; 41-b; 42-a; 43-d; 44-c; 45-e.

Integrative Thinking

1. During fetal life, oxygenated blood is transported via the umbilical vein from the placenta to the fetus. It enters the right atrium of the heart, as all veins ultimately do. From there most of the blood is shunted across the interatrial septum through an opening, the <u>foramen ovale</u>, into the left atrium, thus bypassing the functionless pulmonary circuit and entering the left ventricle and going from there into the systemic circuit. A second bypass is effected by the <u>ductus arteriosus</u>, which shunts what little blood has entered the pulmonary artery across to the thoracic artery and thus to the systemic circuit. At birth, both of these shunts close, and the pulmonary circuit becomes functional. If either of these shunts fails to close at birth, the blood cannot be sufficiently oxygenated, and so

it is mostly the "bluish venous" blood that courses through the systemic circulation, rather than the brilliantly red "arterial" blood.

There is a third cause of blue babies that occasionally occurs. Hemolytic anemia is a condition that may accompany sickle cell anemia, thalassemia, or Rh incompatibility. The fragmented erythrocytes that sometimes characterize these maladies interfere with circulation efficiency to the extent that the blood is insufficiently oxygenated and ,hence, has a bluish tinge.

2. Tell them straight out to avoid all of these, for they are either directly or indirectly responsible for chronic diseases of the circulatory and respiratory systems as well as untimely death. They are also responsible for the soaring costs of health insurance, which everyone must bear. They are especially to be avoided by women during pregnancy, becuse of their deleterious effects on development, particularly on mental development. An extreme example is fetal alcohol syndrome, which too frequently is characterized in the child by empty spaces where parts of the brain should be.